T0331686

UNDERSTANDING
GRAVITY

The Generation Model Approach

Other Related Titles from World Scientific

General Relativity: A First Examination
Second Edition
by Marvin Blecher
ISBN: 978-981-122-043-2
ISBN: 978-981-122-108-8 (pbk)

A Cabinet of Curiosities: The Myth, Magic and Measure of Meteorites
by Martin Beech
ISBN: 978-981-122-491-1

Theory of Groups and Symmetries: Representations of Groups and Lie
Algebras, Applications
by Alexey P Isaev and Valery A Rubakov
ISBN: 978-981-121-740-1

UNDERSTANDING
GRAVITY

The Generation Model Approach

Brian Albert Robson
Australian National University, Australia

World Scientific

NEW JERSEY · LONDON · SINGAPORE · BEIJING · SHANGHAI · HONG KONG · TAIPEI · CHENNAI · TOKYO

Published by

World Scientific Publishing Co. Pte. Ltd.

5 Toh Tuck Link, Singapore 596224

USA office: 27 Warren Street, Suite 401-402, Hackensack, NJ 07601

UK office: 57 Shelton Street, Covent Garden, London WC2H 9HE

Library of Congress Control Number: 2021941148

British Library Cataloguing-in-Publication Data
A catalogue record for this book is available from the British Library.

UNDERSTANDING GRAVITY
The Generation Model Approach

ISBN 978-981-121-491-2 (hardcover)
ISBN 978-981-121-492-9 (ebook for institutions)
ISBN 978-981-121-493-6 (ebook for individuals)

For any available supplementary material, please visit
https://www.worldscientific.com/worldscibooks/10.1142/11683#t=suppl

Desk Editor: Ng Kah Fee

Typeset by Stallion Press
Email: enquiries@stallionpress.com

Printed in Singapore

Preface

> To solve any problem that has never been solved before, you
> have to leave the door to the unknown ajar.
>
> You have to permit the possibility that you do not have it ex-
> actly right.
>
> Otherwise, if you have made up your mind already, you might
> not solve it.
>
> *Richard Feynman*

This book is primarily concerned with the development of a quantum theory
of gravity, which was achieved by the development of an alternative model
of particle physics to the Standard Model (SM) of particle physics, which
I have called the Generation Model (GM) of particle physics. This development took almost two decades from 2002-2019 during which time considerable attention was given to the above statement attributed to Richard
Feynman.

The 20th century experienced enormous progress in physics, essentially
on account of two new theories: relativity theory and quantum mechanics
and the development of new apparatus. Indeed, the nature of the Universe today is described in terms of two important theoretical models, the
Standard Model (SM) of particle physics and the Standard Model of Cosmology (SMC), based upon Einstein's general theory of relativity, which
describes the force of gravity and the large scale structure of the Universe
and quantum mechanics, which describes the physics of the very small.

Unfortunately, as emphasized by Stephen Hawking and others, these two
theories are known to be inconsistent with each other, so that one needs
to accommodate the gravitational force within the domain of quantum

v

mechanics by developing a quantum theory of gravity that will apply for both large and small scales of the Universe.

This incompatibility of the two theories became apparent in the latter part of the 20th century and consequently led to the incompleteness of both the SM and the SMC, together with a general slowing down of progress in the fundamental physics associated with an understanding of the Universe.

Further progress to overcome the incompleteness of the SM was achieved by reviewing several dubious assumptions made during the development of the SM, leading to the development of an alternative model of particle physics, the GM. While the main difference between the GM and the SM is that in the GM the twelve elementary particles of the SM, the six leptons and the six quarks, are composite particles, there are a few additional essential differences, such as the occurrence of mixed quark states in hadrons.

The Introduction to the book outlines the format of the twelve chapters presented in the book. Chapter 1 discusses the Enigma of Gravity. Chapters 2-4 provide a historical perspective of three different eras of physics. Chapters 5 and 6 discuss the development of the SM and the GM, respectively. Chapters 7-11 discuss the many successes of the GM in providing an understanding of many problems and puzzles associated with both the SM and the SMC. Chapter 12 is the Epilogue.

Acknowledgments

Many people have helped in the making of this book. First, several colleagues, by their criticism and constructive comments, arising from reading the many papers and book chapters written during the development of the GM. I wish to record that Neville Fletcher also assisted me during several years prior to his death in 2017 with a joint project to prepare a review paper entitled "Fundamental and Residual Forces". While the SM recognizes four fundamental forces in nature: the gravitational, electromagnetic, weak nuclear and strong nuclear forces, described in almost every modern physics text book, my GM recognizes only two fundamental forces in nature: the electromagnetic and strong nuclear forces. This review paper was intended to be a historical discussion of both classical and quantum forces in nature primarily for the benefit of teachers of physics. Unfortunately, this project was abandoned in 2017, although many of the historical perspectives concerning the nature of forces have since been included in this book. I am also grateful to my philosophical colleague, Kristian van der Pals, who read the whole initial draft of the book, for his many suggestions and critical comments.

Second, I am indebted to Tim Senden, Director of the Research School of Physics in the Australian National University for his support and strong encouragement to complete this book, the members of the Research School of Physics, Computer Unit, for their help in providing assistance for the preparation of the initial draft of the book, and the staff of World Scientific for their assistance in the later stages of production of the book.

Third, I am also indebted to Martinus Veltman for providing the initial inspiration to solve the problem of the three generations of leptons and quarks of the SM and to Walter Greiner for his encouragement and assistance in providing a viable publication journal for non-main-stream ideas.

Finally, I thank Vladimir Kekelidze and Elena Kokoulina from the High Energy Physics Laboratory, Joint Institute for Nuclear Research, Dubna, Russia for inviting me to visit them in 2013 in order to discuss an experiment using their new Nuclotron accelerator to determine the parity of the neutral pion, which was known to be mainly pseudoscalar. An earlier 2008 experiment at Fermi Laboratory had placed a limit on the scalar contribution to the neutral pion decay amplitude of less than 3.3%, while my GM indicated a scalar contribution to the decay amplitude of about 2.5%. Unfortunately, this experiment is difficult so that no lower limit than 3.3% has yet been placed upon the scalar contribution to the decay amplitude.

Contents

Introduction

According to historical records, the quest to understand both the composition and the structure of the Universe has so far lasted for over 2500 years. The nature of the Universe is dependent primarily upon two properties: (1) the nature of the building blocks, i.e. the elementary particles, of the constituent ordinary matter; and (2) the nature of the forces acting between these elementary particles.

In this book, progress in the understanding of the nature of the Universe is divided into three eras: (i) the era of classical physics, which is assumed to run from antiquity (ca. 600 BC) until about 1895 and is associated with the macroscopic world in which only the gravitational and electromagnetic forces are evident from direct experience because of their long-range nature and ordinary matter was considered to be composed of atoms; (ii) the era of transitional physics, 1895-1932, in which several discoveries were made, which could not be reconciled with classical physics and indicated the need for new physics; and (iii) the era of modern physics, 1932 to the present day, which is associated mainly with the microscopic (subatomic) world in which both the weak nuclear and the strong nuclear short-range forces operate.

Currently, the understanding of the nature of the Universe is based primarily upon two important theoretical models: (1) the Standard Model (SM) of particle physics, and (2) the Standard Model of Cosmology (SMC).

Both these models, based primarily upon observations, were essentially completed during the 20th century and were developed employing two new theories, *relativity theory* and *quantum mechanics* that originated in the earlier years of the 20th century.

Today, scientists describe the Universe mainly in terms of two theories: (i) Einstein's general theory of relativity, which describes the force of gravity and the large-scale structure of the Universe; and (ii) quantum mechanics, which describes the physics of the very small. Unfortunately, however, as emphasized by Stephen Hawking and others, these two theories are known to be *inconsistent* with each other, so that one needs to accommodate the gravitational force within the domain of quantum mechanics by developing a *quantum theory of gravity* that will apply for both the large and small scales of the Universe.

The *incompatibility* of the two underlying theories, general relativity and quantum mechanics, means that at least one of these theories is *incomplete*, leading to problems for both the SM and the SMC.

Indeed, the SM is based upon quantum mechanics and although this model recognizes four fundamental forces in nature: the gravitational, electromagnetic, weak nuclear and strong nuclear forces, described in almost every modern physics text book, it does not provide any understanding of the gravitational force, since this force is so much weaker than the other three fundamental forces and is considered to play no role in particle physics.

On the other hand, the SMC is based upon both quantum mechanics for describing the physics of the very small and also the general theory of relativity for describing the physics of the very large in terms of classical physics. Einstein's general theory of relativity is still considered to be the theory that best describes gravity, although in this theory, gravity is not regarded as a force but rather as a consequence of massive objects warping spacetime. However, although Einstein's theory of gravity has been validated by many experiments and observations, providing descriptions of effects that are unexplained by Newton's law of gravity, it suffers not only from being incompatible with quantum mechanics but also from being essentially a mathematical theory, on par with Newton's earlier universal theory of gravitation. I refer to the lack of a physical understanding of Newton's mathematical formula for describing his universal theory of gravity as the *enigma of gravity*.

Chapter 1 discusses this enigma of gravity: What is the physical mechanism underlying the gravitational force? In 1687 Newton announced his universal law of gravitation and in 1915 Einstein announced his new theory of gravitation: in both cases the authors explained with mathematics *how* their laws of gravitation worked but failed to understand the physics, i.e. the *cause* underlying their laws of gravitation.

One of the main aims of this book is to report how an understanding of the enigma of the gravitational force was achieved and led to a quantum theory of gravity.

In 2001, following my attendence at a public lecture presented by Martinus Veltman in Canberra, Australia concerning the "Facts and Mysteries in Elementary Particle Physics", I understood that most physicists considered that the SM is *incomplete* in the sense that it provides no understanding of several empirical observations, including in particular the occurrence of three families or generations of elementary particles of the SM.

Furthermore, this incompleteness of the SM did not necessarily arise from any incompleteness of quantum mechanics upon which the SM is based, but possibly arose from several *dubious* assumptions made during the development of the SM during the 20th century.

The three eras will be discussed from a historical perspective in Chapters 2-4, and will indicate the changing natures of both the building blocks of ordinary matter and the forces acting between these elementary particles.

Chapter 5 discusses the long-term development of the SM throughout the three eras of physics, covered in Chapters 2-4, and summarizes the changing properties of both the elementary particles and the fundamental forces of the SM. This chapter also discusses that the incompleteness of the SM implies a need for an alternative model of particle physics.

Chapter 6 reports my two-decade attempt to develop a satisfactory alternative model to the SM that describes the occurrence of three families or generations of the elementary leptons and quarks of the SM. I called this model, the Generation Model (GM): it was only achieved by removing several *dubious* assumptions inherent in the SM.

I consider that this GM has been remarkably successful in providing an understanding of many problems and puzzles associated with both the SM and the SMC. These are discussed in the following Chapters 7-11.

In Chapter 7, the origin of mass is discussed. Within the framework of the SM, most of the mass of the Universe arises from the internal energy content of the constituent particles of ordinary matter, but a small percentage arises from the Higgs mechanism, providing mass to the elementary (originally massless) leptons and quarks of the SM, so that the origin of mass is not *unified*. On the other hand, the GM provides a *unified* origin of mass: the mass of a body m is a measure of its energy content E and is given by $m = E/c^2$, where c is the speed of light in a vacuum, according to Einstein's 1905 conclusion, deduced from his special theory of relativity.

Chapter 8 discusses the solution to the enigma of gravity: in the GM, gravity is not a fundamental force but is a very weak, universal and attractive complex residual interfermion color force acting between the colorless particles, electrons, neutrons and protons, that essentially constitute the total mass of a body of ordinary matter. This complex residual color force provides a quantum theory of gravity.

Chapter 8 also describes the notions of both 'dark matter' and 'dark energy' that the SMC claims constitute about 27% and 68%, respectively, of the mass-energy content of the Universe. However, both dark matter and dark energy are simply names describing unknown entities, and consequently constitute two very dubious assumptions of the SMC. In the GM both dark matter and dark energy are explained in terms of properties of the new quantum gravity theory.

Chapter 9 discusses the failure of the SM to account for the virtual absence of antimatter in the Universe. This is called the 'matter-antimatter asymmetry problem', since it is generally considered that the present Universe consists almost entirely of matter particles, although it is generally assumed that the Universe was created in the Big Bang from pure energy, which produced equal numbers of particles and antiparticles, i.e. matter and antimatter: Where have all the antiparticles gone? The GM provides a possible solution to this matter-antimatter asymmetry problem, in terms of composite, rather than elementary, leptons and quarks.

Chapter 10 discusses the consequences of the postulate of the GM that hadrons are composed of weak eigenstate quarks rather than mass eigenstate quarks. The main consequences are the existence of higher generation quarks in nucleons and the existence of mixed parity states in hadrons. In particular the mixed parity of the charged pions leads to a quantitative description of the decay of the long-lived K_2^0 meson into two charged pions *without the violation of CP symmetry* in the CC weak nuclear interaction process. This is contrary to the earlier interpretation of the surprising discovery by Cronin, Fitch and collaborators in 1964 of the unexpected decay of the K_2^0 meson into two charged pions.

Chapter 11 discusses the Glashow, Weinberg and Salam (GWS) model of the relation between the electromagnetic and the weak nuclear interactions known as the electroweak connection. It is noted that several problems that arise during the derivation of the electroweak connection within the framework of the SM may be avoided by assuming that the weak nuclear interaction is not a fundamental interaction. Consequently, the electroweak connection may be derived within the framework of the GM assuming a global gauge invariance, rather than a local gauge invariance as in the GWS model.

Chapter 12 is the epilogue.

Chapter 1

The Enigma of Gravity

In about 1666 Isaac Newton (1642-1727) claimed that he had considered why apples in his orchard in Woolsthorpe always fell to the ground. He argued that an apple fell because it was attracted by the Earth's gravitational force. This led him to consider whether the same gravitational force caused the motion of the Earth, planets and comets around the Sun.

During the 17th century several scientists, including Johannes Kepler (1571-1630) and Robert Hooke (1635-1703) had considered that the Earth, planets and comets were kept in orbit around the Sun by some force of attraction. However, they were unable to state the precise nature of this attractive force.

Moreover, in 1609 Kepler had published two laws of planetary motion: (1) the orbit of a planet is an ellipse with the Sun at one of the foci and (2) a line segment joining a planet and the Sun sweeps out equal areas during equal intervals of time. Kepler had obtained these laws by analyzing for some five years astronomical observations of Tycho Brahe (1546-1601). In 1619 Kepler had published a third law: (3) the square of the orbital period of a planet is proportional to the cube of the semi-major axis of its orbit.

In antiquity Greek philosophers such as Aristotle (ca. 384-322 BC) and Archimedes (ca. 287-212 BC) employed the concept of force primarily in the study of stationary and moving objects: a force was considered to be any interaction that, when unopposed, changed the motion of a body of matter. Unfortunately, they believed that a force was necessary to maintain even uniform motion because of an incomplete understanding of friction, air resistance, etc.

Galileo Galilei (1564-1642) corrected this misunderstanding by carrying out an experiment in which bodies of matter were rolled down an inclined plane. He showed that such objects were accelerated by gravity,

1

independently of their mass and concluded that bodies of matter retain their velocity unless acted upon by a force such as friction or air resistance.

Galileo also introduced the fundamental concept of the *relativity of motion*: he recognized that it is impossible to detect uniform motion along a straight line unless such motion was compared relative to some inertial reference frame. An *inertial* frame of reference is one in which an isolated object, experiencing no force, moves along a straight line with uniform velocity. In addition Galileo assumed that both time and the fundamental laws of physics are the same in *all* inertial frames of reference moving with constant velocity with respect to one another. This assumption became known as *Galilean relativity*.

During 1664-66, which included the period of the Great Plague, so that Newton spent most of this time at his family home in Woolsthorpe away from his college, Trinity, in Cambridge, Newton developed a new branch of mathematics, which he called *fluxions* that is now known as *calculus*. Newton also considered the nature of the centripetal force required to maintain a body in a circular orbit: he was very interested in similar forces that keep the planets in their elliptical orbits around the Sun. Indeed he claimed that he had deduced that these forces must vary inversely as the squares of their distances from the Sun. Newton's developments in mathematics would prove to be very useful for him to establish his three laws of motion and a universal law of gravitation.

Later Newton, assuming Galileo's relativity principle and also the existence of a frame of reference of absolute rest, formulated laws of motion that were not improved upon for over 200 years.

Newton's *First Law of Motion* states that objects continue to move in a state of constant velocity unless acted upon by an external net force or resultant force. This is essentially a definition of *inertia*, i.e. the resistance of any physical object to any change in its state of motion, including changes to its speed, direction or state of rest.

Newton's *Second Law of Motion* states that the force $\mathbf{F}$ acting upon an object of mass m will result in the center-of-mass of the body being accelerated by an amount $\mathbf{a}$, i.e. $\mathbf{F} = m\mathbf{a}$. Mass is a property of a physical body that conceptually is difficult to define uniquely. This will be discussed in more detail later.

Newton's *Third Law of Motion* states that whenever a first object exerts a force $\mathbf{F}$ on a second object, the second object exerts a force $-\mathbf{F}$ on the first object. This means that in a closed system of objects, there are no internal forces that are unbalanced.

Newton's three laws of motion are the same in all inertial frames of reference, in agreement with Galilean relativity, and essentially defined the classical mechanical age of physics.

In 1687 Newton announced both his three laws of motion and his universal theory of gravity in his book, "The Mathematical Principles of Natural Philosophy" *Philosophiae Naturalis Principia Mathematica*, generally known as the *Principia*, considered by many physicists to be the most influential physics book ever written. After considerable contemplation he postulated a *universal law of gravitation*: the force on a spherical body of mass m_1 due to the gravitational force of another spherical body of mass m_2 is given by: $\mathbf{F} = -Gm_1m_2\mathbf{r}/r^3$, where $\mathbf{r}$ is the radial vector pointing in the direction away from the center of mass of the first body toward the center of mass of the second body and G is Newton's gravitational constant.

Newton showed that according to his universal law of gravitation the planets, which are approximately spherical bodies, move in elliptical orbits around the Sun in agreement with Kepler's three laws.

Newton's universal law of gravitation simply states *how* the gravitational force of attraction acts between any two spherical bodies: it increases in direct proportion to the product of their masses and decreases in inverse proportion to the square of their distance apart. However, Newton's theory of gravity suggests no physical mechanism to explain *why* the universal law of gravitation provides a good description of many natural phenomena.

Newton placed more importance upon the requirement that his law of universal gravitation should describe the observations rather than speculations about the *why* of the gravitational force, e.g. why was the attraction proportional to the product of the two masses of the interacting bodies and why did the attraction diminish according to the inverse-square law?

Newton emphasized that his universal law of gravitation relied upon his mathematical principles, especially his universal laws of motion, which in turn depended entirely upon observations. He declared that *'Hypotheses non fingo'*, i.e. 'I do not make hypotheses'. In particular, he also declared that he could not understand the *cause* of the gravitational force. This lack of understanding, which to a large extent remains the case today, concerning the physical causes describing Newton's universal law of gravity, is responsible for the *enigma* associated with the gravitational force. Newton had explained the mathematics rather than the physics of the phenomena described by his universal law of gravitation.

The main aim of this book is to report how an understanding of the enigma of the gravitational force was achieved. To some extent, a solution of this enigma was discovered accidently as a by-product of the solution of another problem. The solution is primarily based upon an understanding of the nature of the various forces that have been found to occur between the various elementary particles of a physical system composed of *ordinary* matter.

According to the prevailing Standard Model of Cosmology (SMC) [1], the Universe is composed of about 5% ordinary matter, 27% dark matter and 68% dark energy. The definition of ordinary matter is usually given in terms of its *building blocks* or *elementary particles* and consequently its definition has changed considerably throughout history.

In antiquity the building blocks of ordinary matter were called *atoms* from the Greek word $\alpha\tau o\mu o\sigma$ meaning *indivisible* and the notion of dark matter did not exist. Today the building blocks of ordinary matter are considered to be the *elementary leptons and quarks* of the Standard Model (SM) of particle physics [2–4]. Both the notions of *dark matter* and *dark energy* will be discussed later: currently the SM provides no understanding of either of these dubious assumptions of the SMC.

Currently the elementary particles of the SM are assumed to interact via four fundamental forces [3, 4]: (1) gravitational, (2) electromagnetic, (3) weak nuclear and (4) strong nuclear. These four forces have different properties:

(1) the *gravitational force* acts between any two bodies of ordinary matter since mass or energy is its source; it is an extremely weak, attractive and long-range force;

(2) the *electromagnetic force* acts between bodies of ordinary matter carrying electric charge and can be attractive or repulsive depending whether the interacting electric charges are opposite in sign or the same sign; this force includes the electrostatic force acting between electrically charged particles at rest and the combined effect of electric and magnetic forces acting between electrically charged particles moving relative to each other;

(3) the *weak nuclear force* is responsible for radioactive decay and neutrino interactions; it has a very short range, is very weak and is the only known force that does not conserve parity;

(4) the *strong nuclear force* acts between bodies of ordinary matter carrying *color* charge and can be attractive or repulsive depending upon the nature of the color charges; it has a finite range, approximately the size of a nucleon.

In April 2001 I attended a public lecture presented in Canberra, Australia by a then recent Nobel laureate, Martinus Veltman (1931-2021) concerning the "Facts and Mysteries in Elementary Particle Physics", prior to the publication in 2003 of his book with the same title [3]. Veltman stated that the *greatest puzzle of elementary particle physics was the occurrence of three families of elementary particles that have the same properties except for mass* in the SM. A solution of the enigma of the gravitational force was achieved by initially solving the occurrence of the three families, later termed *generations*, of elementary particles of the SM by the development of an alternative model to the SM, which I called the Generation Model (GM) [5].

This book will indicate that the gravitational force is a universal attractive very weak complex residual force of the strong nuclear force, acting between the relevant building blocks of the GM associated with the three massive particles, the proton, the neutron and the electron, which are the constituents of a body of ordinary matter. In addition, this gravitational force of the GM has two important properties that differ from the Newtonian gravitational force, which provide an understanding of both dark matter and dark energy.

Chapter 2

Era of Classical Physics

2.1 Introduction

In this chapter and several subsequent chapters, a historical introduction into the development of an understanding of both the *building blocks*. i.e. the various elementary particles, of ordinary matter and the *nature of the forces* occurring between these elementary particles, will be presented.

Chapter 2 will describe the era of classical physics associated with the macroscopic world in which only the *gravitational* and *electromagnetic* forces are evident from direct everyday experience because of their long-range nature and ordinary matter was considered to be composed of discrete units called *atoms*. The era of classical physics will be assumed to run from antiquity (ca. 600 BC) until about 1895, prior to the development of two theories *relativity theory* and *quantum mechanics* that ultimately led to further significant progress in the understanding of the nature of both ordinary matter and forces.

Chapter 3 will describe the era of transitional physics in which during the years 1895-1900 several discoveries were made, which could not be reconciled with classical physics and indicated the need for new physics [3,4,6,7]. This chapter will discuss both the *special theory of relativity* proposed by Albert Einstein (1879-1955), which replaced the earlier Galilean relativity of Newtonian mechanics, leading to the *general theory of relativity* (Einstein's theory of gravity) and the *quantum hypothesis* introduced into physics by Max Planck (1858-1947), which eventually led to the *quantum mechanics* of Werner Heisenberg (1901-1976) and others in 1925. The era of transitional physics covers the years 1895 to 1932.

Chapter 4 will describe the era of modern physics associated mainly with the microscopic (subatomic) world in which both the weak nuclear and the strong nuclear short-range forces operate. The era of modern physics will be assumed to have been initiated in 1932.

2.2 Matter and Atomism

In antiquity both the Greek philosophers, Leucippus (ca. 480-420 BC) and his pupil Democritus (ca. 460-370 BC) considered that ordinary matter is composed of discrete units, which they named *atoms* from the Greek word ατομοσ meaning *indivisible*. However, it took about two millenia before the atomic theory of ordinary matter was established in 1803 by John Dalton (1766-1844) as a fundamental theory for chemical reactions.

Dalton's atomic theory assumed (1) all matter is made of atoms that are indestructable and indivisible; (2) all atoms of a particular element have the same mass and chemical properties; (3) atoms of different elements have different masses and different chemical properties; (4) chemical compounds consist of two or more different kinds of atoms and (5) a chemical reaction is a rearrangement of atoms.

Dalton's atomic theory remains essentially valid today for chemical reactions so that Dalton's theory provides the theoretical foundation of chemistry. However, in the modern era, we know that while atoms cannot be destroyed by chemical reactions, atoms can be destroyed by nuclear reactions. Furthermore, there are atoms of a given element with different masses that are known as *isotopes*, although isotopes of an element have the same chemical properties.

2.3 Electromagnetic Force

The gravitational force is not the only force known in the macroscopic or classical world. The electromagnetic force has also been known from the time of ancient Greece. Indeed, the word *electromagnetism* is derived from combining two Greek terms: *electron* for amber and *magnetis lithos* for magnesian stone. Thales (ca. 624-546 BC) was aware of magnetic materials (lodestones) and also that electric charge could be generated by rubbing fur on amber.

Two kinds of electric charge were discovered in 1733 by Charles-François de Cisternay du Fay (1698-1739), which he named *vitreous* and *resinous* (later known as positive and negative) electric charge, respectively. He found these two kinds of electric charge by rubbing glass with silk to obtain vitreous electric charge and by rubbing amber with fur to obtain resinous electric charge. He also found that like-charged objects repel each other while unlike-charged objects attract one another.

The electrostatic force between two static electrically charged particles with charges q_1 and q_2 was discovered in 1785 by Charles Augustin de Coulomb (1736-1806). This electrostatic force is given by Coulomb's law: $\mathbf{F} = k_e q_1 q_2 \mathbf{r}/r^3$ where $\mathbf{r}$ is the radial vector pointing away from charge q_1 and towards charge q_2 and k_e is Coulomb's constant. The force $\mathbf{F}$ acting on q_1 by q_2 is attractive (repulsive) if $q_1 q_2$ is negative (positive) for unlike (like) charges, respectively.

If the two electric charges are not static but are *moving*, additional forces come into play: these are called magnetic forces. Evidence for such forces was discovered in antiquity: Thales and others noticed that 'lodestones', naturally magnetized pieces of the mineral magnetite, could attract iron. The word *magnet* is derived from the Latin word *magnetum* for lodestone and this in turn was derived from the Greek words meaning a stone from Magnesia, a part of ancient Greece.

In 1269 Peter Peregrinus de Maricourt wrote an important article *Epistola de magnete* concerning the fundamental laws of magnetism. In particular he introduced the notion of magnetic *poles* and discussed the existence of two poles of a magnet. He indicated how a magnet may act as a *compass* with the two poles being termed the North and South poles, depending whether the pole is oriented approximately towards the Earth's North magnetic pole or South magnetic pole, respectively. He also stated that like poles repel each other while unlike poles attract each other. Peregrinus attributed the Earth's magnetism to the action of celestial poles such as the pole star, rather than to terrestrial poles of the Earth itself. However, in 1600 William Gilbert (1544-1603) in his book *de Magnete* concluded that the center of the Earth was iron and that the Earth itself behaved like a magnet.

The study of magnetism developed separately from that of electricity until Hans Christian Oersted (1777-1851), in 1820, noted that a compass needle is deflected from magnetic North when an electric current from a nearby battery is switched on or off. Shortly afterward, Andre-Marie Ampere (1775-1836) confirmed Oersted's observation and demonstrated that a coil of wire carrying an electric current behaves like an ordinary magnet. He also showed that two parallel wires carrying electric currents attract or repel each other, depending on whether the currents flow in the same or opposite directions, respectively. Subsequently, he derived Ampere's law describing more generally the magnetic force between two electric currents.

In 1831 Michael Faraday (1791-1867) found that if a magnet is moved through a loop of wire that an electric current flowed in the wire. He then employed this technique to construct the electric dynamo, the first electric power generator.

In 1838 Faraday proposed that electromagnetic forces extended into empty space around a conductor of electricity. His concept of *lines of force* emanating from charged bodies and magnets led to the idea of electric and magnetic fields. This constituted a paradigm shift in the concept of a force: in Newton's time no field was associated with the gravitational force so that no one thought of there existing something in the space between the planets and the Sun. It should be noted that an essential property of a *field* is that it contains *energy*: to create either a magnetic field by passing an electric current along a thin wire or to create an electric field in a loop of wire by moving a magnet through the loop requires energy.

In 1865 James Clerk Maxwell (1831-1879) published his landmark paper [8] in which a set of twenty partial differential equations describe how the electric and magnetic fields vary in space due to sources, electric charges and electric currents or magnets, respectively. These equations demonstrated that electric and magnetic forces are two complementary aspects of electromagnetism. In addition, Maxwell calculated that an electromagnetic field could propagate through space as a wave moving with the speed of light. This led him to suggest that light is an electromagnetic disturbance propagating in space according to electromagnetic laws.

In 1884 Oliver Heaviside (1850-1925), employing vector calculus, reduced twelve of Maxwell's original twenty equations in twenty unknowns down to four differential equations in four unknowns that are now known as *Maxwell's equations*. These equations describe the nature of electric fields, magnetic fields and the relationship between the two electromagnetic fields.

In 1889, Heaviside derived the magnetic force on a moving charged particle. Finally, in 1892, Hendrik Lorentz (1853-1928) derived the modern form of the electromagnetic force, which includes contributions from both the electric and magnetic fields. This force is known as the Lorentz force: $\mathbf{F} = q(\mathbf{E} + \mathbf{v} \times \mathbf{B})$, where $\mathbf{E}$ and $\mathbf{B}$ are the electric and magnetic fields, respectively, acting upon a particle with charge q and velocity $\mathbf{v}$.

In the era of classical physics prior to 1900, only two of the conventionally accepted fundamental forces, the *gravitational* force and the *electromagnetic* force are evident from direct experience because of their long-range nature. At the beginning of the 20th century, both these forces were considered to be *fundamental* forces, although in the case of the electromagnetic force, its nature changed significantly from the idea of objects exerting a force upon each other to the notion of a *field* containing energy acting between the two objects.

The fundamental *gravitational* force was considered to act between any two spherical bodies of masses, m_1 and m_2, respectively, according to Newton's universal law of gravitation: $\mathbf{F} = -(Gm_1m_2\mathbf{r})/r^3$, where $\mathbf{r}$ is the radial vector pointing in the direction away from the center of mass of the first body toward the center of mass of the second body and G is Newton's gravitational constant. The accepted value of G at the end of the 19th century was 6.66×10^{-11} m^3.kg^{-1}.s^{-2} with an uncertainty of 0.2%.

The fundamental *electromagnetic* force was considered to be the Lorentz force: $\mathbf{F} = q(\mathbf{E} + \mathbf{v} \times \mathbf{B})$, where $\mathbf{E}$ and $\mathbf{B}$ are the electric and magnetic fields, respectively, acting upon a particle with charge q and velocity $\mathbf{v}$.

Unlike the *gravitational* force, there were several *residual electromagnetic* forces recognized at the beginning of the 20th century. These *residual* forces were of two main kinds: (1) *intramolecular* forces and (2) *intermolecular* forces.

An *intramolecular* force was any force that held together the atoms comprising a molecule. Several types of *intramolecular* forces were recognized: ionic, covalent and metallic, each involving *fundamental* electromagnetic forces between the electrically charged constituents of the electrically neutral atoms comprising the molecule.

An *intermolecular* force was a force, which acts between atoms, molecules or ions. *Intermolecular* forces were weak compared to *intramolecular* forces but also involved *fundamental* electromagnetic forces. Examples were dipole-dipole forces, hydrogen bonding and van der Waals forces.

Without knowledge of the internal structure of atoms, Coulomb forces were the only interatomic forces well understood in the 19th century. Prior to the development of quantum mechanics in the 1920s, both the intramolecular and intermolecular forces were treated phenomenologically, especially using a hard-sphere model of the atom [9].

Lorentz regarded electric charge as a local deficiency or surplus of small charged particles, later named electrons, hidden within atoms, while electric currents consisted of moving electrons. Electrons caused electromagnetic fields, which in turn acted upon stationary or moving charged particles. This view of electromagnetism is presented in Lorentz's book "The Theory of Electrons" [10].

2.4   Luminiferous Aether

In 1865 Maxwell described light as an electromagnetic field that propagated through space as a wave moving with the speed of light. This suggested that light was travelling through an electromagnetic medium that pervades all space termed the *luminiferous aether*. In addition, Maxwell's equations required that all electromagnetic waves propagate in vacuum at a fixed speed c.

Furthermore, Maxwell's equations implied that the aether represented an absolute and unique reference frame in Newtonian mechanics. Maxwell suggested that detecting motion relative to the aether should be possible, since under a Galilean transformation the equations of Newtonian mechanics are invariant whereas those of electromagnetism are not.

An experiment to determine the velocity of the Earth through the aether was carried out in 1887 by Albert Michelson (1852-1931) and Edward Morley (1838-1924). They obtained a *null* result, which was the first clear demonstration that something was seriously wrong with the aether hypothesis.

One explanation of the null result of the Michelson-Morley experiment was that it showed that the aether surrounding the Earth moves with the velocity of the Earth itself. In order to rescue the stationary aether hypothesis, George FitzGerald (1851-1901) in 1889 and Lorentz in 1892, independently proposed a remarkable hypothesis, now known as the *Lorentz-FitzGerald contraction*, in which a moving object's length is measured to be shorter than the length measured in the object's own rest frame. Since this contraction hypothesis was considered to be *ad hoc* another explanation of the null result was sought.

This was achieved in 1905 by Einstein who introduced a new relativity theory, later called the *special theory of relativity* and demonstrated that the contraction hypothesis did not require motion through a stationary aether [11]. This special theory of relativity will be discussed in more detail in the following Chapter 3.

Chapter 3

Era of Transitional Physics

3.1 Introduction

Prior to 1895 atoms were still considered to be the basic units of matter. However, several discoveries: X-rays, radioactivity and the electron, were made during the years 1895-1900, which could not be reconciled with classical physics and indicated the need for new physics [3, 4, 6, 7].

In this chapter the era of transitional physics, 1895-1932, will be discussed. Two theories, relativity theory and quantum mechanics, were initiated in the earlier years of the 20th century that ultimately led to further significant progress in the understanding of the nature of both matter and forces.

Often the development of new apparatus provides progress in science. One important device was a glass tube named a Crookes tube after William Crookes (1832-1919) in the 1870s. This glass tube was evacuated to very low pressure and a high electrical potential of thousands of volts was applied between an anode and a cathode placed inside the tube. It was found that the glass behind the anode *glowed*, due to particles (initially known as cathode rays) emitted from the cathode causing the glass to fluoresce.

X-rays were discovered in 1895 by Wilhelm Roentgen (1845-1923) while investigating cathode rays in a high vacuum Crookes tube. Roentgen found that some invisible rays emanating from the tube could pass through books and papers on his desk. Later he discovered their medical use by making a photograph of his wife's hand.

Another important experimental development was the cloud chamber invented by Charles Wilson (1869-1959) in 1911. He found that tiny water droplets tended to form around charged ions in vapour-saturated air, so that an electrically charged particle, e.g. an electron, leaves a trail of small droplets as it traverses the cloud chamber.

In 1879 Crookes claimed that cathode rays were a stream of negatively charged particles. During the years 1894-1897 John J. Thomson (1856-1940) also studied the phenomenon of cathode rays. By measuring the deflection of the cathode rays in combined electric and magnetic fields, Thomson was able to determine both the speed of the particles and the ratio e/m, where e is the electric charge and m is the mass of each particle. He also showed that these particles were identical to those particles emitted from various metals by ultraviolet light (the photoelectric effect) by repeating such earlier experiments. Using a Wilson cloud chamber, Thomson also estimated an approximate value for the electric charge e of the cathode ray particles and found that the mass of these particles was about 1800 times smaller than the mass of the hydrogen atom. For these investigations, Thomson was credited as the discoverer of the first *subatomic* particle later named the *electron*.

The first accurate measurement of the electron charge was made by Robert Millikan (1868-1953). This determined the electron mass to be 1/1840 of the mass of the hydrogen atom.

In 1896 Henri Becquerel (1852-1908) discovered accidently the phenomenon of radioactivity, while investigating phosphorescent substances, since he suspected that the glow occurring in Crookes tubes by X-rays might be associated with phosphorescence. Becquerel had wrapped a photographic plate in black paper, had placed a uranium salt upon it and had stored the combination within a drawer for several days. He found that the uranium salt emitted a mysterious radiation, which had caused a blackening of the photographic plate during the few days that it had been within the drawer. This radiation process was named *radioactivity* in 1898 by Pierre (1859-1906) and Marie (1867-1934) Curie following their discovery of a new element, named *polonium* after the country of Marie's birth, as a result of their investigations of different radioactive substances. Also in 1898 another radioactive substance, named *radium* was discovered by the Curies assisted by Gustave Bémont (1857-1932). The discovery of radium launched the use of radium for the treatment of cancer and the commencement of modern nuclear medicine.

In 1898 Ernest Rutherford (1871-1937) made an important discovery. By wrapping a sample of uranium in successive layers of aluminium foil, Rutherford showed that the uranium radiation is complex and that there are present at least two distinct kinds of radiation: one that is readily absorbed and the other of a more penetrative character. These different radioactive radiations were later termed alpha (α) and beta (β) rays, respectively. Subsequently, it was determined that the α-radiation consists of helium nuclei particles called α-particles while the β-radiation consists of electrons. In 1900 Paul Villard (1860-1934) discovered a third kind of radiation emanating from radium, γ-rays. These were later shown to be identical to X-rays.

In 1909 Hans Geiger (1882-1945) and Ernest Marsden (1889-1970) investigated the scattering of α-particles from a metal plate, following a suggestion by Rutherford. They found that about 1 in 8000 α-particles were reflected, i.e. were scattered by more than 90°. Rutherford's reaction to this back-scattering effect was to exclaim that '*it was almost as incredible as if you fired a 15-inch shell at a piece of tissue paper and it came back and hit you*'. Subsequently, Rutherford carried out a series of calculations based upon the results of the α-particle scattering experiments and concluded that most of the mass of an atom resided within a minute *nucleus* that had a positive charge, equal in magnitude ($+Ze$) to the total electric charge ($-Ze$) of all the electrons in the neutral atom, and a diameter only $1/100000$ of that of the atom as a whole. Consequently he determined that most of the atom is empty space. The integer Z is known as the *atomic number* of the element.

In 1919 Rutherford employed sufficiently high energy α-particles to bombard N^{14} forming O^{17} with the emission of a hydrogen nucleus. Similar experiments with a variety of light nuclei also caused a hydrogen nucleus to be emitted. Consequently, it was concluded that the hydrogen nucleus was one of the building blocks of all other nuclei and it was accorded a special name: the *proton* from the Greek word $\pi\rho o\tau o\sigma$, meaning first. In this way Rutherford discovered the proton, only the second subatomic particle to be discovered.

During the years 1911-1912 Victor Hess (1883-1964) discovered that a continuous flux of low-intensity radiation existed in the upper atmosphere and that it originated outside of the Earth's atmosphere. This radiation was later named *cosmic rays* and studies indicated that the radiation consisted of fast charged particles that had energies considerably greater than those resulting from the decay of radioactive nuclei.

Subsequently, it was determined that cosmic rays provided a source of fast charged particles, which were sufficiently energetic that these particles could produce new particles. Many new particles were discovered during studies of cosmic rays in the years 1932-1952, prior to the development of powerful accelerators that could produce high energy charged particles, such as protons: such accelerators could produce these new particles in a more controlled manner. These discoveries will be discussed in the following Chapter 4.

In the 1920s physicists faced a serious problem with the β-rays found in radioactive radiation. The energy spectra of the electrons emitted in the radioactive decay of certain nuclei consisted of a continuous curve rather than a discrete energy as was found for α-rays. This discrete energy corresponded to the difference between the mass of the original nucleus and the mass of the resultant alpha particle and the remaining nucleus. By analogy with the α-rays, β-rays appeared to violate the basic conservation laws of physics.

In 1930 Wolfgang Pauli (1900-1958) proposed a solution to the above problem: in a β-decay an as yet undetected particle, later called the *electron antineutrino* (see Chapter 5) carries off the appropriate energy so that energy is conserved. The proof of Pauli's proposal was provided in 1956 by Clyde Cowan (1919-1974) and Frederick Reines (1918-1998), who studied the reverse process to β-decay emitted from a nuclear reactor, and successfully identified the existence of the proposed particle.

The atomic masses of various elements, A, had been measured during the 19th century employing chemical reactions. More accurate measurements were used in the early 20th century by finding the e/m ratios, using vacuum tubes, e.g. in 1913 Thomson measured the ratio e/m for ions of neon ($Z = 10$) and found that there were two kinds of neon atoms with $A = 20$ (Ne^{20}) and $A = 22$ (Ne^{22}), respectively. The earlier measurements, based upon chemical reactions that did not distinguish between the two kinds of atoms, found that the mass of the neon atom was $A = 20.18$. Thomson showed that this value was simply the weighted average of the two kinds of neon atoms. Thus it was found that atoms could have identical nuclear charge (Z) but different atomic masses (A). Such atoms were termed *isotopes* of the same element from the Greek words $\iota\sigma o$ meaning equal and $\tau o\pi o\sigma$ meaning place.

Atomic masses are measured in atomic mass units (amu) and one amu is defined to be 1/12 of the mass of the common carbon atom (C^{12}). Measurements showed that the atomic mass of each isotope, when expressed in amu, is always quite close to an integer. This implied that the nuclei of all the elements are composed of a small number of building blocks.

It was clear that the nuclei of all the other elements, other than H^1, are not composed of protons only, otherwise Z would be always equal to A. It was eventually suggested that there might be an electrically neutral particle, which gave the atomic nucleus the rest of its mass. This hypothesis was confirmed in 1932 when the *neutron*, with a mass close to that of the proton, was discovered by James Chadwick (1891-1974). The neutral neutron was the third subatomic particle to be discovered.

In the 1930s the conventional atom was considered to consist of a minute nucleus, composed of Z protons and A-Z neutrons, surrounded by a cloud of Z electrons. Thus the *proton, neutron and electron were now the elementary particles of matter.*

3.2 Relativity Theory

In the early years of the 20th century, two theories, relativity theory and quantum mechanics, were initiated that ultimately led to further significant progress in the understanding of the nature of both matter and forces. First the progress in relativity theory will be discussed.

As indicated in the previous chapter, in 1905 Einstein introduced a new relativity theory, the *special theory of relativity*. This relativity theory is based upon two assumptions: (1) the speed of light in a vacuum is the *same* for all inertial observers, i.e. those moving with uniform velocity, and is independent of the motion of the light source; (2) the laws of physics are the *same* in all inertial frames of reference.

Einstein's special theory of relativity implies the replacement of the earlier Galilean relativity of Newtonian mechanics, based upon both an absolute time and a stationary inertial frame of reference, and defined by the Galilean transformations:

$$x' = x - vt, \quad y' = y, \quad z' = z, \quad t' = t, \tag{3.1}$$

by transformations (later termed *Lorentz transformations*):

$$x' = \beta(x - vt), \quad y' = y, \quad z' = z, \quad t' = \beta(t - vx/c^2), \tag{3.2}$$

where $\beta = 1/(1 - v^2/c^2)^{1/2}$ and v is the relative uniform velocity along the x-axis.

The Lorentz transformations, Eq. (3.2), describe the Lorentz-FitzGerald contraction, since each of the observers ascribes a contraction to the length measurement along the x-axis made by the other observer.

The special theory of relativity implies several additional consequences, associated with the nature of forces, including the variation of mass with velocity and the equivalence of mass and energy. These have been experimentally verified.

Newtonian mechanics assumes that the mass of a particle is *constant* under all conditions of velocity. Special relativity implies that the mass m of a particle in an inertial frame of reference moving with relative uniform velocity v is given by:

$$E = mc^2 = m_0 c^2 (1 - v^2/c^2)^{1/2} = \beta m_0 c^2 , \qquad (3.3)$$

where E is the energy and m_0 is the *rest mass* of the particle.

This relativistic variation of mass leads to an advance of the perihelion of an elliptical orbit associated with an inverse square force law, such as assumed in both Newton's universal law of gravity and Coulomb's law of the electrostatic force. The former provides a contribution to the advance of the perihelion of the planet Mercury, while the latter describes a contribution to the *fine structure* of the hydrogen atom.

The Lorentz transformations of the special theory of relativity also imply that the speed of light represents an upper limit for the speed at which any physical interaction may be transmitted. This led Einstein to note that the special theory of relativity therefore *conflicted* with Newton's universal law of gravitation, since Newton's law implied that the gravitational interaction acted *instantaneously* for cosmological distances.

Consequently, Einstein searched for a theory that possessed all the successful features of Newton's universal law of gravitation but did *not* conflict with the special theory of relativity. After considerable contemplation between 1905 and 1915, Einstein developed a new theory of gravity named the *general theory of relativity* [12–14].

In 1915 the final version of general relativity was formulated: in this theory, gravity is not regarded as a force but rather as a consequence of massive objects warping spacetime, which describes the three dimensions of space together with the one dimension of time [12]. Currently, general relativity is generally considered to be the theory that best describes gravity. This new theory of gravity proposed by Einstein in 1915 was his generalization of the special theory of relativity to include accelerating frames of reference in addition to inertial frames of reference.

In Einstein's theory of gravity, spacetime is no longer Euclidean geometry but a different geometry based upon the Karl Schwarzschild (1873-1916) solution of the equations of general relativity. Essentially, Einstein substituted the concept of curved spacetime for the mysterious Newtonian action at a distance.

The solution found by Schwarzschild in 1916 was an *exact* solution of Einstein's equations and provided the curved spacetime geometry surrounding a spherical mass. This solution, which is consistent with the special theory of relativity, predicts an advance of the perihelion of an elliptical orbit associated with an inverse square force law as assumed in Newton's universal law of gravity.

The observed advance of the perihelion of the planet Mercury is about 574 arc-seconds per century, which Urbain Le Verrier (1811-1877) calculated in 1859 to be about 40 arc-seconds per century larger than the total perturbation caused by all the known other planets. This observation was the first indication that Newton's universal law of gravitation did not completely explain the slow precession of Mercury's orbit around the Sun. Indeed, the anomalous precession of 43 arc-seconds per century was eventually described by Einstein using the Schwarzschild solution of his general theory of relativity [12].

This early success of general relativity inspired confidence in Einstein's new theory of gravity. However, it was the observation of the deflection of light rays near the Sun during a total solar eclipse that established the viability of Einstein's theory.

According to the Schwarzschild solution, the deflection of light rays near the Sun was calculated to be about 1.76 arc-seconds, which is exactly *twice* the value predicted by Newton's theory of gravity, if photons, i.e. quanta of energy, are subject to Newtonian gravitation in the same manner as massive particles.

In 1919 Arthur Eddington (1882-1942), the leader of an expedition to test Einstein's theory of gravity by observing the deflection of light rays from stars passing near the Sun during a total solar eclipse, concluded that the expedition's observations favored Einstein's theory rather than Newton's, although the observations were not sufficiently accurate to be conclusive. However, later measurements using microwaves rather than visible light have confirmed Eddington's conclusion.

Einstein's and Newton's theories of gravitation predict very similar results for the solar system, although the two theories are based upon very different assumptions. In the two examples in the solar system: (1) the precession of the orbit of Mercury and (2) the deflection of light rays by the Sun, the differences in the predictions are small. In hindsight this close agreement appears to arise because Einstein based his gravity theory upon the assumption that Newton's theory was accurate for *weak* gravitational fields. As will be indicated later, this assumption will be shown to be invalid (see Section 8.2), leading to an understanding of why general relativity is unable to describe observations on cosmological scales.

In 1917 Einstein applied the general theory of relativity to constructing a model of the Universe. Initially Einstein based his model on assumptions corresponding to those of Newton's much earlier attempt: the Universe was infinite and the distribution of matter was homogeneous and isotropic on sufficiently large scales. However, in 1929 Edwin Hubble (1889-1953) discovered that light from remote galaxies was redshifted and that the fainter the galaxy the larger was its redshift. Hubble, reluctantly, assumed that the redshift of a galaxy was due to a Doppler effect, implying that the galaxy was moving away from the Earth with a speed that increases with distance. Both Einstein and Newton had assumed that the Universe was essentially *static* so that Hubble's observations implied that the Universe was *not* static but was *expanding*.

In 1922 Alexander Friedmann (1888-1925), assuming that the distribution of matter in the Universe was homogeneous and isotropic, showed that one should not expect the Universe to be *static* according to the general theory of relativity: Friedmann predicted what Hubble discovered later.

In 1927 Georges Lemaître (1894-1966) noted that while the equations of general relativity permitted models of the Universe that were homogeneous and isotropic distributions of matter, these models were not static. In particular, expanding universes could be extrapolated backwards in time to an originating singular point that became associated with the notion of the 'Big Bang'. Lemaître called this original very small and compact system the 'primordial atom' and considered that the present Universe arose as a result of the observed expansion. The prevailing model of the Big Bang, which forms part of the Standard Model of Cosmology (SMC), is based upon the general theory of relativity, which forms one of the pillars of the SMC.

It should be noted that in 1929 Fritz Zwicky (1898-1974) proposed the 'tired light' hypothesis as an alternative interpretation to that of the Doppler effect to explain the distance-redshift relation. Zwicky proposed that the cosmological redshift is caused by the energy loss of photons by their interaction with material particles on their journey through cosmological space. Zwicky's hypothesis was not accepted by most astronomers and cosmologists at the time nor since, although the tired-light hypothesis has recently been considered as a viable alternative to the Big Bang scenario (see Section 4.4).

3.3 Quantum Mechanics

In the mid-1890s Planck started to study a problem associated with radiant heat. The classical view was that the wavelengths of radiant heat given off by a hot body must consist of all possible frequencies. According to the known laws of classical physics, a hot body was supposed to radiate energy of all possible frequencies, up to a definite maximum, depending upon how hot the body was. It was known from experiment that shorter wavelengths of electromagnetic energy were hotter than the longer wavelengths. This implied that as a body became hotter that it would emit an increasing amount of radiant energy: this became known as the *ultraviolet catastrophy problem.*

In 1900 Planck succeeded in solving the ultraviolet catastrophy problem by introducing a *quantum hypothesis*: the absorption and emission of radiation takes place by the transfer of *energy quanta*, i.e. finite elements of energy according to $\Delta E = h\nu$, where $h = 6.626 \times 10^{-34}$ J.s is Planck's constant and ν is the frequency of the radiation. This quantum hypothesis, which was in direct conflict with the well established principles of classical physics, initiated the development of a quantum theory that provided a basis for understanding the subatomic world.

In 1905 Einstein accepted the quantum hypothesis of Planck for the processes of emission and absorption of electromagnetic radiation but also proposed that the radiation itself consists of energy quanta, later called *photons*. Consequently, Einstein's proposal provided not only an explanation of Planck's quantum hypothesis required to solve the ultraviolet catastrophy problem but also provided an explanation of the *photoelectric effect.*

The photoelectric effect is the emission of electrons when substances, particularly metals, are irradiated with light of a wavelength shorter than a certain limiting value. Experiments have indicated that the velocity of the emerging electrons does not depend upon the intensity of light but only depends upon the frequency ν of the incident photons. Einstein essentially proposed that the energy E of the emerging electrons was given by:

$$E = h\nu - B, \tag{3.4}$$

where B is a constant characteristic of the metal, corresponding to the amount of energy required to remove an electron from the metal. Every photon striking the metal and colliding with one of its electrons transfers its whole energy $h\nu$ to the electron so that the energy of the emergent electron is given by Eq. (3.4) according to the conservation of energy.

Since quantum mechanics considers that electromagnetic waves consist of photons, it implies that the electromagnetic field suggested by Faraday is replaced by photons. Thus, in quantum mechanics, the original concept of a force makes little sense. Instead one has *interactions* with charged electrons emitting or absorbing photons, and the concept of a force has become the exchange of a particle.

In 1913 Niels Bohr (1885-1962) applied the quantum hypothesis to the structure of the atom: he inferred that the atom can exist only in definite *discrete stationary states* with energies E_0, E_1, $E_2, \ldots$, so that only those radiation spectral lines can be absorbed for which $h\nu$ has the exact value to raise the atom from one stationary state to a higher one. Bohr assumed, for the hydrogen atom, that the orbits of the electron about the proton corresponded to those predicted classically but which fulfilled certain quantum conditions involving integral *quantum numbers*, corresponding to the stationary states of the atom. In this way, Bohr was able to explain the formula found in 1885 by Johann Balmer (1825-1898) for the discrete atomic hydrogen spectrum:

$$\nu = R(1/n^2 - 1/m^2), \tag{3.5}$$

where $n = 1, 2, 3, \ldots$, and $m > n$ is an integer. $R = 109678$ cm^{-1} is the Rydberg constant.

Soon after 1913, when quantum dynamics was initiated, the first steps were made to unify special relativity and quantum theory. In particular, the Balmer formula given by Eq. (3.5) does not explain that every spectral line, predicted by that formula, is actually a close doublet of lines, called *fine structure*.

In 1914-15 Arnold Sommerfeld (1868-1951) explained the fine structure of the discrete atomic hydrogen spectrum by taking into account the relativistic variation of the electron mass arising from the electron's velocity. Sommerfeld showed that the classical orbits of an electron in the field of a nuclear charge Ze are approximate ellipses, which exhibit a perihelion precession: the fine structure occurs because the relativistic effects lift the degeneracy in the quantized orbital angular momentum characteristic of the non-relativistic theory of states in a Coulomb $1/r$-potential.

Quantum numbers are the values of quantized quantities in the dynamics of a quantum system. For example, in the hydrogen atom, four quantum numbers describe completely the quantized dynamics of the electron.

The *principal* quantum number n (or m) in Eq. (3.5) describes quantized energy levels of the electron. The *azimuthal* quantum number l describes the magnitude of the quantized orbital angular momentum of the electron: the value of l ranges from 0 to $n-1$; $l=0$ is called an s orbital, $l=1$ a p orbital, $l=2$ a d orbital, etc. Thus for $n=1$ we have only $l=0$, i.e. a 1s orbital; for $n=2$ we have $l=0$ and 1, i.e. 2s and 2p orbitals; and for $n=3$ we have $l=0$, 1 and 2, i.e. 3s, 3p and 3d orbitals. The *magnetic* quantum number m_l describes the quantized orbit that yields a projection of the quantized orbital angular momentum along a specified axis: the values of m_l range from $-l$ to $+l$ with integer intervals. Thus for $l=0$ we have $m_l = 0$; for $l=1$ we have $m_l = -1$, 0 and $+1$; for $l=2$ we have $m_l = -2$, $-1, 0, +1$, and $+2$; etc. The *spin projection* quantum number m_s describes the quantized *intrinsic* spin angular momentum of the electron within a quantized orbital and gives the projection of the quantized spin angular momentum along the specified axis. The only values of m_s are $-\frac{1}{2}$ and $+\frac{1}{2}$, since the electron has intrinsic spin $s = \frac{1}{2}$.

The mathematical expression derived in 1911 by Rutherford for the scattering of α-particles allowed the value of Z of the scattering atom to be determined. It was found that if the elements are arranged in the order of the *periodic table*, initiated by Dimitri Mendeleev (1834-1907), that their values of Z increase in consecutive numbers so that hydrogen has $Z=1$, helium has $Z=2$, lithium has $Z=3$, etc. In addition, elements in the same vertical column in the periodic table have similar chemical properties. This *periodicity* implied some connection between chemical properties and the number of electrons in the atom.

In 1925, in order to account for the total number of electron orbitals in an atom, Pauli introduced the *exclusion principle*, which states that no two electrons can exist in the same quantum state defined by the above four quantum numbers, n, l, m_l and m_s.

In 1926 both Enrico Fermi (1901-1954) and Paul Dirac (1902-1984) independently showed that the Pauli exclusion principle applied to identical particles with *half-integer* spin in a system with thermodynamic equilibrium. Such particles are known as *fermions* because Fermi published first. Dirac also pointed out that the Pauli exclusion principle did *not* apply to identical particles with *integer* spin. Such particles are known as *bosons*, named after Satyendra Bose (1894-1974), who investigated them.

The difference between fermions and bosons arises since fermions obey Fermi-Dirac quantum statistics, which describe a system of identical *half-integer spin* particles, e.g. electrons in thermal equilibrium, by *antisymmetric* many-particle wave functions, while bosons obey Bose-Einstein quantum statistics, which describe a system of identical *integer spin* particles, e.g. photons in thermal equilibrium, by *symmetric* many-particle wave functions.

In 1925 Heisenberg stated that the basic reason for the failure of the Bohr quantum theory is that it deals with quantities which are *unobservable*, so that the fundamental ideas of Bohr's theory can never be tested. Heisenberg said that in order to develop a consistent system of atomic physics, later termed *quantum mechanics*, only observable entities should be introduced into the theory, e.g. the frequencies and intensities of light emitted by an atom rather than the orbits of electrons. Such a system termed *matrix mechanics* was developed by Heisenberg in collaboration with Max Born (1882-1970) and Pascual Jordan (1902-1980) in 1925.

In 1926 Erwin Schrödinger (1887-1961) developed a second version of quantum mechanics based upon the wave nature of particles, according to the hypothesis of Louis de Broglie (1892-1987) that to every particle there corresponds a wave, the wave length λ of which is connected to the momentum p of the particle by the relation $\lambda = h/p$, involving Planck's constant h. This version is called *wave mechanics*.

In 1927 Dirac showed that matrix mechanics and wave mechanics were essentially equivalent using his so-called *transformation theory*, which related the linear operators of wave mechanics to the corresponding matrix operators of matrix mechanics. In both mechanics the quantum allowed values of each operator, representing a given physical quantity, are the eigenvalues of the operator [15].

In 1927 Heisenberg introduced his so-called 'uncertainty principle' asserting that certain pairs of physical properties of a particle, e.g. *position* and *momentum* cannot both be determined precisely at the same time. Another pair are *energy* and *time*, which implies that a particle may have an energy that does not correspond to its actual momentum, provided that this occurs only for a short time in agreement with the uncertainty relation relating energy and time. This is important for the nature of the interactions between particles (see Chapter 4).

Heisenberg's 'uncertainty principle' is essentially a consequence of the dualistic nature of particles, arising from the two versions of quantum mechanics: in wave mechanics, particles have 'wave properties' such as a wave length λ that is related to its 'particle property' momentum p by the relation $\lambda = h/p$. This relation leads to Heisenberg's uncertainty relationship, which states that the product of the uncertainties in determining position (Δx) and momentum (Δp) is approximately equal to Planck's constant h, i.e. $\Delta x \cdot \Delta p \approx h$. Similarly, the wave property, frequency ν, is related to the particle property, energy, by the relation $E = h\nu$. This leads to Heisenberg's second uncertainty relationship, which states that the ratio of the uncertainties in determining energy (ΔE) and frequency $(\Delta \nu)$ is approximately equal to h: i.e. $\Delta E / \Delta \nu \equiv \Delta E \Delta t \approx h$. Heisenberg's uncertainty relations completely overthrew the determinism of classical Newtonian mechanics.

Unfortunately, Schrödinger's quantum mechanics did not include the special theory of relativity, so that it was not guaranteed to work at speeds close to the speed of light. However, the unification of quantum mechanics and special relativity was accomplished in 1928 by Dirac, who derived a relativistic wave equation called the *Dirac equation*.

In 1926 Oskar Klein (1894-1977) and Walter Gordon (1893-1940) proposed the Klein-Gordon equation to describe relativistic electrons. However, this equation does not account for the spin of the electron but simply describes spinless (scalar) relativistic particles.

In 1927 Charles Galton Darwin (1887-1962) suggested that the electron should be described by a two-component wave equation in order to take into account the intrinsic spin of the electron.

Shortly afterwards Pauli introduced the intrinsic angular momentum of the electron in a fixed direction as a new quantized quantity. Since the electron has spin $\frac{1}{2}$, this fixed direction can only take two values, $-\frac{1}{2}$ and $+\frac{1}{2}$. Consequently, Pauli introduced a realization of the spin angular momentum in terms of a vector $\mathbf{s}$, which he represented by 2×2 matrices. The components of σ, where $\mathbf{s} = (\hbar/2)\sigma$ are given by the Pauli matrices [6, 15, 16]:

$$\sigma_1 = \begin{pmatrix} 0 & 1 \\ 1 & 0 \end{pmatrix}, \quad \sigma_2 = \begin{pmatrix} 0 & -i \\ i & 0 \end{pmatrix}, \quad \sigma_3 = \begin{pmatrix} 1 & 0 \\ 0 & -1 \end{pmatrix}, \quad (3.6)$$

so that in the fixed direction 3:

$$s_3 \begin{pmatrix} \psi_1 \\ \psi_2 \end{pmatrix} = (\hbar/2) \begin{pmatrix} \psi_1 \\ -\psi_2 \end{pmatrix}, \quad (3.7)$$

and ψ_1 and ψ_2 are eigenfunctions of spin corresponding to eigenvalues, $(+\hbar/2)$ and $(-\hbar/2)$, respectively. Here $\hbar = h/2\pi$ and is approximately 10^{-34} J.s. However, Pauli realized that while his two-component form of Schrödinger's equation did take into account the intrinsic spin of the electron, it did not conform with special relativity.

In 1927 Dirac considered that the relativistic equation for the electron should be *first order* in both time and the three space variables, in order to be invariant under Lorentz transformations.

Dirac was intrigued by the equation:

$$(\sigma \cdot \mathbf{p})^2 = (\mathbf{p}^2 \cdot 1_2), \qquad (3.8)$$

where 1_2 is the 2×2 unit matrix. He considered how to generalize Eq. (3.8) so as to factorize the sum of *four* squares rather than three, required for the relativistic generalization of Eq. (3.8):

$$(p_1)^2 + (p_2)^2 + (p_3)^2 - (E/c)^2 = -(mc)^2, \qquad (3.9)$$

involved in the relativistic Klein-Gordon equation for a free particle of mass m [17]:

$$[(E/c)^2 - (p_1)^2 - (p_2)^2 - (p_3)^2 - (mc)^2]\psi(x_1, x_2, x_3, t) = 0. \qquad (3.10)$$

Dirac obtained the linearization of Eq. (3.10) by assuming the required equation to be of the form:

$$[(E/c) - \alpha_1 p_1 - \alpha_2 p_2 - \alpha_3 p_3 - \alpha_0 mc]\psi(x_1, x_2, x_3, t) = 0, \qquad (3.11)$$

where α_0, α_1, α_2 and α_3 are 4×4 matrices that satisfy the relations:

$$(\alpha_\mu)^2 = 1_4 = \begin{pmatrix} 1 & 0 & 0 & 0 \\ 0 & 1 & 0 & 0 \\ 0 & 0 & 1 & 0 \\ 0 & 0 & 0 & 1 \end{pmatrix}, \qquad (3.12)$$

for $\mu = 0,1,2,3$ and

$$\alpha_\mu \alpha_\nu + \alpha_\nu \alpha_\mu = 0_4 = \begin{pmatrix} 0 & 0 & 0 & 0 \\ 0 & 0 & 0 & 0 \\ 0 & 0 & 0 & 0 \\ 0 & 0 & 0 & 0 \end{pmatrix}, \qquad (3.13)$$

for $\mu \neq \nu$ and $\mu, \nu = 0,1,2,3$.

Dirac introduced the following 4×4 matrices as the simplest matrices that satisfy the relations (3.12) and (3.13):

$$\alpha_0 = \begin{pmatrix} 1_2 & 0_2 \\ 0_2 & -1_2 \end{pmatrix}, \quad \alpha_1 = \begin{pmatrix} 0_2 & \sigma_1 \\ \sigma_1 & 0_2 \end{pmatrix}, \qquad (3.14)$$

$$\alpha_2 = \begin{pmatrix} 0_2 & \sigma_2 \\ \sigma_2 & 0_2 \end{pmatrix}, \quad \alpha_3 = \begin{pmatrix} 0_2 & \sigma_3 \\ \sigma_3 & 0_2 \end{pmatrix},$$

where

$$1_2 = \begin{pmatrix} 1 & 0 \\ 0 & 1 \end{pmatrix}, \quad 0_2 = \begin{pmatrix} 0 & 0 \\ 0 & 0 \end{pmatrix}, \qquad (3.15)$$

and σ_1, σ_2 and σ_3 are the Pauli matrices of Eq. (3.6).

When the 4×4 matrices are introduced into Eq. (3.11), the wavefunction ψ accordingly becomes a column vector with four components:

$$\psi = \begin{pmatrix} \psi_1 \\ \psi_2 \\ \psi_3 \\ \psi_4 \end{pmatrix}, \qquad (3.16)$$

and thus Eq. (3.11) becomes a set of four simultaneous equations known as the *Dirac equation*.

In 1928 Dirac's formulation of a four-component relativistic wave equation Eq. (3.11) describing an electron, provided a *unification* of quantum mechanics with special relativity. Dirac also found that his equation successfully predicted all the electron properties resulting from its spin: spin angular momentum and its magnetic moment. However, the physical meaning of *four* components was not immediately obvious: in 1928 there were only *three* known particles, the electron, the proton and the photon.

In 1931, following considerable contemplation, Dirac concluded that his four-component Eq. (3.11) not only described the spin properties of the electron but also described the existence of an associated particle with the same mass but with opposite (positive) charge of the electron that Dirac called the *antielectron*.

In 1932 Carl Anderson (1905-1991) discovered a positively charged particle with the same mass as the electron while using a cloud chamber to study cosmic ray particles. Anderson employed a cloud chamber in which a magnetic field was applied. The magnetic field causes the path of a charged particle to curve, the curvature of a positively charged particle being opposite to that of a negatively charged particle. In order to determine the direction of the charged particle, Anderson placed a lead plate in the cloud chamber. Since the particle lost some kinetic energy traversing the lead plate, the charge of the particle could be determined from knowledge of its direction of travel and the curvature of its path. Anderson identified several examples of a positively charged particle with a mass comparable to the electron and called them *positrons*. Anderson's observations were carried out independently of the Dirac theory so that the discovery of the positron was essentially accidental.

The positron was the first *antiparticle* to be discovered. Dirac's equation predicts that for every charged fermion there exists an antiparticle with the same mass but with the opposite charge, e.g. the proton p^+ has an antiparticle p^-, the antiproton discovered in 1955 by Owen Chamberlain (1920-2006) and Emilio Segrè (1905-1989). It turns out that every fermion, including neutral fermions such as the neutron have corresponding antiparticles having the same mass as the particle but *opposite* values of their *intrinsic additive quantum numbers* including charge (see Chapter 4). It has become conventional to represent an antiparticle by a 'bar' above the symbol representing the particle, e.g. the antiproton is represented by $\bar{p}^+$ or simply $\bar{p}$, rather than p^-.

In 1928 Dirac *unified* quantum mechanics with the special theory of relativity. However, Einstein's general theory of relativity has *not* yet been unified with quantum mechanics. Currently, most physicists consider that the general theory of relativity and quantum mechanics are *incompatible* and that general relativity needs to be replaced by a quantum theory of gravity [18]. Such a theory is discussed in Chapter 8.

Although during the era of transitional physics, quantum mechanics proved to be extremely successful in solving many problems in both physics and chemistry, phenomena were discovered in 1947 that defied it. First, Willis Lamb (1913-2008) and Robert Retherford (1912-1981) measured a small but finite energy gap between the 2s and 2p eigenstates of the hydrogen atom. This was entirely unexpected, since Dirac's equation predicted that these two energy levels should have the same energy. Second, Polykarp Kusch (1911-1993) measured that the electron magnetic moment was slightly larger than the value predicted by Dirac's equation. It was soon realized that these phenomena required a reformulation of relativistic particle quantum mechanics.

This led to the development of a relativistic quantum field theory of the interaction of photons with electrons. In quantum field theory, the photons are not the primary reality but correspond to excitations of the electromagnetic field associated with electrons. The primary reality status of the electromagnetic field and the association of its excitations with the quantized electrons describes the *indistinguishability* of all electrons. Thus this initial field theory consisted of a relativistic quantum field of electromagnetic radiation (photons), a relativistic quantum field of matter (electrons) that acts as the source of electromagnetism, and the interactions between the radiation and matter fields. Although this theory led to agreement with the experimental result of Kusch for the electron magnetic moment, it was quickly discovered that it yielded an *infinite* result for the Lamb shift.

This infinity problem was overcome by three physicists: Sin'itirō Tomonaga (1906-1979), Julian Schwinger (1918-1994) and Richard Feynman (1918-1988), who independently, using different mathematical methods, managed to 'renormalize' the theory to remove the infinities arising in the calculations.

Without going into the complex mathematical details of relativistic quantum field calculations [3], the discussion here will briefly describe the *renormalization process* initiated in 1947 by Hendrik Kramers (1894-1952) to remove the infinities arising in the calculations.

Kramers considered that the mass of an electron resulted partly from the electric field energy surrounding the electron so that one must clearly separate the 'bare' mass, i.e. the mass not including the field contribution, and the 'physical' mass, that one observes experimentally. Kramers suggested that since the bare mass is not observable, one should choose it so that after inclusion of the field contribution, one obtains the observed value of the mass.

This *renormalized* quantum field theory is called Quantum ElectroDynamics (QED). It was the first such theory to be discovered and is now an integral part of the SM (see Chapter 5). Indeed, QED is by far the most accurate theory in all of science: it now provides a calculation of the Lamb shift, which agrees with the measured value, to 12 significant figures.

The main impact of quantum mechanics upon the concepts of ordinary matter and forces is three-fold. First, the unification of the special theory of relativity and quantum mechanics indicated that matter consists of *both* particles and antiparticles. Second, quantum mechanics led to Faraday's original concept of the electromagnetic force as arising from the existence of an electromagnetic *field* within the space between interacting charged bodies, to the notion of photons being exchanged between the charged bodies. While physicists have considered that the gravitational force may arise similarly from the exchange of a particle known as the *graviton*, the existence of such a particle has not yet been confirmed by experiment. Third, the observation of phenomena that disagreed with the predictions of the Dirac equation led to the development of a renormalizable relativistic quantum field theory called QED, which describes the interaction of photons with electrons and positrons.

Chapter 4

Era of Modern Physics

4.1 Introduction

Chapter 4 will describe the era of modern physics associated mainly with the microscopic (atomic) world in which both the weak nuclear and the strong nuclear short-range forces operate.

The year 1932 was a very important year in which several significant discoveries and advances were made in the understanding of the nature of both ordinary matter and forces, and is the year I associate with the initiation of the era of modern physics.

Anderson had discovered a positively charged particle with the same mass as the electron, the positron, which is the antielectron contemplated by Dirac to exist according to his four-component relativistic equation.

Chadwick had discovered the neutron as a constituent of atomic nuclei so that an atom was considered to consist of a minute nucleus composed of Z protons and $(A - Z)$ neutrons surrounded by a cloud of Z electrons. Thus atoms were no longer the basic building block of ordinary matter, being replaced by the three subatomic particles, the *electron*, the *proton* and the *neutron*.

The discovery of the neutron raised two important questions: (1) What holds the neutrons and protons together within the atomic nuclei? and (2) What force provides an explanation of the continuous energy spectra observed in β-decay radiations?

In this chapter I shall discuss the above question (1), which will involve the concept of a *strong nuclear force*: the stability of an atomic nucleus, composed of many neutrons and protons, requires the existence of a new type of force that is stronger than the electromagnetic repulsion between the constituent protons.

In 1932 Heisenberg, assuming that atomic nuclei are composed of neutrons and protons, commenced developing models to describe their structures in terms of the various nuclear forces acting between the atomic constituents. In particular he introduced the notion of *strong isospin*.

The discovery of the neutron also implied that there are no electrons within atomic nuclei so that any theory of β-decay is required to account for the process whereby a neutron is converted into a proton, an electron and an electron antineutrino (see Chapter 3). This necessitated the existence of a second nuclear force (later termed the *charge-changing (CC) weak nuclear force*).

4.2   Strong Isospin

In 1932 Heisenberg suggested that the proton and the neutron, which had very similar masses and appeared to be subject to the same nuclear force, could be regarded as two quantum states of the same particle that he called the *nucleon*. By analogy with ordinary spin, Heisenberg considered the nucleon to have *strong isospin*, $\mathbf{I} = \frac{1}{2}$, with the two values of its strong isospin projection quantum number, $I_3 = \pm\frac{1}{2}$, corresponding to the proton and neutron, respectively.

Heisenberg realized that the approximate equality between the number of protons and neutrons in atomic nuclei, especially for light nuclei, e.g. C^{12}, implied that the strong nuclear force was *short-ranged* and also that the strong nuclear force between any two nucleons, i.e. $n-n$, $n-p$ and $p-p$ strong nuclear forces, were very similar, essentially charge independent.

This nuclear symmetry, described by the concept of strong isospin, provided an understanding of nuclear isobars, i.e. nuclei having the same atomic masses (A). Thus, if electromagnetic forces were neglected, nuclei such as Li^7 and Be^7 would be identical. Similarly, the proton and neutron would become identical and have equal masses.

These considerations led to the notion of *mass multiplets*, i.e. systems having approximately the same mass but different charges, with a relation between the charge Q and the strong isospin projection quantum number I_3. Thus for a general isobaric nucleus, assuming $I_3 = +\frac{1}{2}$ for each proton and $I_3 = -\frac{1}{2}$ for each neutron, one has

$$Q = I_3 + \frac{1}{2}A \,, \tag{4.1}$$

where I_3 is the sum of the isospin projection quantum numbers of the A constituent protons and neutrons.

Equation (4.1) relates the charge Q to the strong isospin projection quantum number I_3 and the atomic mass A. All these three quantum numbers are known as *additive quantum numbers* since each represents the sum of the corresponding additive quantum numbers of all the particles comprising the composite isobaric nucleus.

The atomic mass A corresponds to the *baryon number*, introduced by Ernest Stückelberg (1905-1984) in 1938 to account for the stability of ordinary matter. Stückelberg proposed that if the proton and the neutron are assigned baryon number $A = +1$, while the photon, electron, positron, electron antineutrino and electron neutrino have $A = 0$, conservation of baryon number forbids the decay of the proton into a positron and other neutral particles. In addition conservation of baryon number forbids the decay of the neutron into an electron and a positron, although it does allow the decay of a neutron into a proton, an electron and an electron antineutrino as in β-radiation radioactivity.

Both the charge Q and the baryon number A quantum numbers have been found to be *conserved* in all reactions occurring in nature. This means that in an equation describing an interaction between particles that the sums of both charges and baryon numbers on the left hand side of an equation are identical with the sums on the right hand side. For example, noting that for each particle additive quantum number N, the corresponding antiparticle has the additive quantum number $-N$, the reaction:

$$p + p \rightarrow p + p + n + \bar{n}, \tag{4.2}$$

is allowed and is observed, whereas the reaction:

$$p + p \rightarrow p + p + n, \tag{4.3}$$

is forbidden and is not observed.

In 1935 Hideki Yukawa (1907-1981) published his theory concerning the nature of the strong nuclear force that binds the nucleons within the nucleus to one another. Following Heisenberg's earlier notions of the strong nuclear force, Yukawa assumed that the strong forces between any two nucleons were both *short-ranged* and *very similar*. With these assumptions, Yukawa estimated that the mediating particles (later called *pions*) should exist in the three charge states, $Q = +1$, 0 and -1, and should have masses intermediate between the electron and the proton.

Following the discovery of the neutron in 1932 by Chadwick, employing energetic α-particles from polonium nuclei to accomplish the nuclear reaction:

$$\alpha + \text{Be}^9 \rightarrow \text{C}^{12} + n, \tag{4.4}$$

all new particles were discovered thereafter either in cosmic rays or with accelerators or reactors, e.g. in 1932 Anderson discovered the positron while investigating the nature of cosmic rays.

In 1936 Anderson and Seth Neddermeyer (1907-1988), while studying cosmic rays, discovered a new particle that has a mass intermediate between the mass of an electron and the mass of a proton. Initially, this particle was considered to be Yukawa's predicted strong nuclear force mediating particle and it was called a *meson* after the Greek word $\mu\epsilon\sigma\varsigma$ for intermediate. However, by 1947 this new particle was found to have the wrong properties to be recognized as the predicted Yukawa particle, since it behaved more like the electron, existing in both positive and negative charge states. It became known as the *muon* (μ). The muon did not interact strongly with atomic nuclei, as was expected, if it was the predicted Yukawa particle.

In 1947 Donald Perkins (1925-) found an event in cosmic rays in which a particle (later termed the π-meson or simply the *pion*) interacted strongly with a nucleus, indicating that the particle was one of Yukawa's predicted mediating particles of the strong nuclear force.

Later in 1947, Cecil Powell (1903-1969) and his group, using photographic emulsion techniques to study cosmic rays, found two events demonstrating the decay of a pion into a muon plus a neutral particle, later determined to be a neutrino-like particle.

The discovery of a strongly interacting meson of the type predicted by Yukawa led to an extension of the strong isospin concept to this particle called the pion. Since Yukawa's theory of the strong nuclear force required three charge states of the pion, it was allotted strong isospin $\mathbf{I} = 1$ and since it was also assigned baryon number $A = 0$, Eq. (4.1) is satisfied for the three values of $I_3 = +1$, 0 and -1, corresponding to the three charge states $Q = +1$, 0 and -1, respectively.

The discovery of the antiproton ($\bar{p}$) in 1955 by Chamberlain and Segrè employing the interaction:

$$p + p \rightarrow p + p + p + \bar{p}, \qquad (4.5)$$

and the discovery of the antineutron ($\bar{n}$) in 1956 using the interaction:

$$p + \bar{p} \rightarrow n + \bar{n}, \qquad (4.6)$$

confirmed that the antiproton ($Q = -1$) and the antineutron ($Q = 0$) satisfy Eq. (4.1) with $A = -1$ and $I_3 = -\frac{1}{2}$ and $I_3 = +\frac{1}{2}$, respectively. In the interactions given by Eq. (4.5) and Eq. (4.6) both the strong isospin projection quantum number and the baryon number are conserved.

To summarize: the concept of strong isospin is very useful for understanding phenomenologically strongly interacting processes involving nucleons, pions and antinucleons. Indeed the concept has been extensively employed in nuclear physics.

Rather than studying cosmic rays to observe new particles by chance, physicists preferred to develop accelerators that could produce high energy charged particles, such as protons, in a more controlled manner. The first proton accelerator was constructed by John Cockcroft (1897-1967) and Ernest Walton (1903-1995) in 1932. This accelerator was used to carry out the first test of the relationship:

$$m = E/c^2 , \qquad (4.7)$$

which Einstein had concluded in 1905, from the theory of special relativity, described the mass of a body m in terms of its energy content E and the speed c of light in a vacuum. Using the nuclear reaction:

$$p + \text{Li}^7 \rightarrow \alpha + \alpha + 17.2 \text{ MeV}, \qquad (4.8)$$

Cockcroft and Walton found that the decrease in mass in the disintegration process (often referred to as *splitting the atom*) was consistent with the observed release of energy according to Eq. (4.7). The energy that an electron gains from an electric field of one volt is known as an electron volt (eV).

In the late 1940s there were several events observed in cosmic rays that indicated the existence of more new particles. In 1947 George Rochester (1908-2001) and Clifford Butler (1922-1999) discovered an important unusual event in their cloud chamber pictures. This event showed a forked track, which they interpreted as the spontaneous decay of a neutral particle into a pair of charged particles. This neutral particle, called a V-particle (now known as a *kaon*), was estimated to have a mass intermediate between that of a pion and that of a nucleon.

Kaons have provided important information about the nature of interactions, since their discovery in cosmic rays. However, significant progress concerning the nature of interactions was only realized when powerful accelerators of various types essentially provided a more systematic approach to the production of new particles such as kaons.

The Brookhaven Cosmotron [7] was the first particle accelerator, specifically a proton synchrotron, to impact kinetic energy in the range of GeV (10^9 eV) to a single particle, accelerating protons to 3.3 GeV. By 1952 this accelerator was producing a significant number of V-particles. It was found that the new particles behaved very strangely: they were produced copiously, but decayed very slowly.

The accelerator experiments demonstrated that the new particles were produced in pairs (termed associated production), a typical reaction being:

$$\pi^- + p^+ \to \Lambda^0 + K^0 \,, \tag{4.9}$$

but decayed individually in about 10^{-10} s. Such a mean lifetime is about 10^{12} times longer than expected if the production and decay mechanism are governed by the same interaction. In the interaction (4.9), K^0 is a neutral kaon, while Λ^0 is a *hyperon* having a mass between those of the neutron and the deuteron, according to the nomenclature adopted at the Bagnères-de-Bigorre conference in 1953.

The paradox of these *strange* particles was resolved by the introduction of a new additive quantum number called *strangeness* (S) by Murray Gell-Mann (1929-2019), Abraham Pais (1918-2000) and Kazuhiko Nishijima (1926-2009). Strangeness was assumed to be conserved in strong nuclear interactions but not necessarily so in weak nuclear interactions. Thus strange particles were produced copiously in pairs via a strong nuclear interaction but decayed individually very slowly via a CC weak nuclear interaction (see Section 4.3).

The introduction of a new additive quantum number, S, which was conserved in both strong and electromagnetic interactions led to a search for a higher symmetry, which was an extension of the strong isospin concept, based upon an $SU(2)$ symmetry.

In 1961 Gell-Mann and Yuval Ne'eman (1925-2006) independently proposed a new model for classifying *hadrons*, i.e. particles influenced by the strong nuclear force, based upon an $SU(3)$ symmetry. The name hadron is based upon the Greek word for strong: $\alpha\delta\rho\circ\varsigma$. This model, later named the *eightfold way* [19], considered the division of the hadrons into 'families' comprising several multiplets, i.e. those having the same baryon number and strong isospin number, satisfying relation (4.1), into a larger set called a *supermultiplet*, satisfying the following relation, including the strangeness quantum number S:

$$Q = I_3 + \frac{1}{2}(A + S)\,. \tag{4.10}$$

The supermultiplets were also postulated to have the same spin angular momentum.

Gell-Mann and Ne'eman proposed that the Λ, Σ and Ξ hyperons, discovered during the 1950s with accelerators, together with the nucleons form an octet of an $SU(3)$ symmetry. These six strange hyperons were assumed to be *displaced* strong isospin multiplets according to Eq. (4.10) with $\mathbf{I} = 0$ for the Λ^0; $\mathbf{I} = 1$ for the Σ^+, Σ^0, Σ^-; and $\mathbf{I} = \frac{1}{2}$ for the Ξ^0, Ξ^-; with the new additive quantum number, strangeness, having the values: $S = -1$ for both Λ and Σ hyperons and $S = -2$ for the Ξ hyperons. The nucleons making up the octet of spin-$\frac{1}{2}$ baryons have, of course, $S = 0$.

Similarly, the strange kaons, also discovered in the 1950s, together with the three pions and another meson, η^0, discovered in 1961, were proposed to form another octet with $SU(3)$ symmetry. The four kaons were assumed to be *displaced* strong isospin multiplets according to Eq. (4.10) with $\mathbf{I} = \frac{1}{2}$ for both the (K^+, K^0) and $(\bar{K}^0, K^-)$ pairs; $\mathbf{I} = 1$ for π^+, π^0, π^-; and $\mathbf{I} = 0$ for the η^0, with the strangeness quantum number having the values: $S = +1$ for K^+ and K^0; $S = -1$ for $\bar{K}^0$ and K^-; and $S = 0$ for the pions and the η^0 meson. This octet consisted of eight spin-0 mesons.

The observed $SU(3)$ symmetry led to a search for an understanding in terms of its *fundamental* representation, corresponding to a triplet of particles. Following several attempts, the *quark model* was proposed independently in 1964 by Gell-Mann and George Zweig (1937-). The members of the fundamental $SU(3)$ triplet were assumed to be three kinds of spin-$\frac{1}{2}$ particles called up (u), down (d) and strange (s) *quarks* by Gell-Mann (Zweig's equivalent particles were called 'aces', since Zweig considered that eventually there would be *four* fundamental particles), consisting of a strong isospin doublet (u, d) and a strong isospin singlet (s). Each quark is assumed to have $A = \frac{1}{3}$, since three quarks comprise a baryon. The u-quark and the d-quark, where u and d stand for the 'up' and 'down' direction of the strong isospin projection quantum number, have $Q = +\frac{2}{3}$, $I_3 = +\frac{1}{2}$ and $Q = -\frac{1}{3}$, $I_3 = -\frac{1}{2}$, respectively and $S = 0$, while the s-quark stands for the strange quark and has $Q = -\frac{1}{3}$, $I_3 = 0$ and $S = -1$. These quark quantum numbers satisfy the charge relation Eq. (4.10).

The 1964 quark model [3, 4, 6, 7] considered that all the hadrons known in the 1960s were composed of the three basic quarks and their three basic antiquarks. Each meson was composed of a quark-antiquark pair, e.g. the π^+ meson was a (u, $\bar{d}$) pair, while each baryon was composed of three quarks, e.g. the proton was a (u, u, d) triplet, and each antibaryon was composed of three antiquarks, e.g. the antiproton was a ($\bar{u}$, $\bar{u}$, $\bar{d}$) triplet. Thus mesons have integral spins, 0 or 1, while baryons and antibaryons have half-integral spins, $\frac{1}{2}$ or $\frac{3}{2}$.

To summarize: the 1964 three quark model provided an excellent description of all the various known hadrons at that time, in terms of their

$SU(3)$ properties. In particular, the model explained both the strong isospin and the strangeness quantum numbers: a hadron's strong isospin projection quantum number was determined by the number of up and the number of down quarks in it and a hadron's strangeness corresponded with the number of strange quarks in it. However, the quark model raised new concerns.

The main problem arose because the quark model indicated that several particles were predicted to have *three identical quarks* in the same quantum state. In the spin-$\frac{3}{2}$ baryon decuplet [3], the Δ^{++} member, composed of three u-quarks, has strong isospin $\mathbf{I} = \frac{3}{2}$, spin angular momentum $J = \frac{3}{2}$ and $S = 0$. Furthermore, the multiplet appeared to behave as an S-wave state of three quarks so that it was *symmetric* in spin. Thus the multiplet seemed to violate the Pauli exclusion principle.

The above problem was resolved in 1965 by Yoichiro Nambu (1921-2015) and Moo-Young Han (1934-2016), who introduced a new degree of freedom for each quark. The quarks were allotted an additional quantum number called *color*, which can take three values so that in effect there are three kinds of each quark, u, d and s. Nowadays, the quarks are considered to carry a *single color charge*: red, green or blue, and the corresponding antiquarks are considered to carry a *single anticolor charge*, antired, antigreen or antiblue: color charge is somewhat analogous to electric charge, although it is associated with an $SU(3)$ symmetry rather than a $U(1)$ symmetry.

Han and Nambu also introduced the notion that the color degree of freedom was associated with a new 'color' symmetry, $SU(3)_C$, and that the quarks interacted via eight vector bosons (later called *gluons*), which acted as an octet in $SU(3)_C$ but as singlets in $SU(3)_f$: the original $SU(3)$ symmetry became known as 'flavor' $SU(3)_f$ symmetry. In addition they proposed that the lowest mass hadrons were $SU(3)_C$ *singlets*: the baryons being composites of three quarks, each having a different color, while the mesons were composites of a quark and an antiquark of opposite colors.

The nature of the gluon fields is such that they lead to a 'runaway growth' of the fields surrounding an isolated color charge [20]. In fact all this structure implies that an isolated quark would have an infinite energy associated with it. This is the reason why *isolated* quarks are not observed. Nature requires these infinities to be essentially cancelled or at least made finite.

It does this for hadrons in two ways: either by bringing an antiquark close to a quark, i.e. forming a meson, or by bringing three quarks, one of each color, together, forming a baryon, so that in each case the composite hadron is *colorless*. It should be noted that in the color charge theory of Han and Nambu that the combination of the three different colors, red + green + blue, is equivalent to the combination of two opposite colors, red + antired, i.e. 'zero' color charge. However, quantum mechanics prevents the quark and antiquark of opposite colors or the three quarks of different colors from being placed *exactly* at the same place. This means that the color fields are *not* exactly cancelled, although sufficiently it seems to remove the infinities associated with isolated quarks.

Currently, the strong nuclear force is considered to arise between quarks carrying a color charge, red, green or blue, and consequently is different in character from the force between colorless hadrons. The force between hadrons is a *residual interaction* acting between all the colored quarks of one hadron and all the colored quarks of the other hadron. This residual interaction is still sufficiently strong so that the neutrons and protons are bound together within atomic nuclei. The mediating particles between the colorless hadrons are colorless mesons such as the pions.

In 1973 the force between particles carrying a color charge, which has been termed the *chromodynamic force*, was developed into a more complete theory called *Quantum ChromoDynamics* (QCD), after the Greek word $\chi\rho o\mu\alpha$ for color by Harald Fritzsch (1943-), Gell-Mann and Heinrich Leutwyler (1938-). The theory assumes that the chromodynamic force is mediated by eight electrically neutral massless particles having spin-1, called gluons. Each gluon carries both a single color charge and a single anticolor charge, and consequently gluons exert chromodynamic forces upon each other. These so-called 'self-interactions' of the gluons lead to two important consequences: (1) *antiscreening effects* leading to an increase in the strong nuclear force field as the separation between the quarks increases and (2) *color confinement* leading to a finite range of the strong nuclear force [4].

In 1964 the existence of a fourth quark was predicted by Sheldon Glashow (1932-) and James Bjorken (1934-), even though three quarks, the up, down and strange quarks, explained all the hadrons known during the 1960s. Prior to the 1964 quark model, it was known that the hadrons were very different from the leptons, being comprised of quarks and very numerous, while the leptons were very few in number. The 1964 quark model implied that the basic building blocks of ordinary matter were the *leptons and quarks*. These two kinds of building blocks were very similar in that they were assumed to be *point* particles, i.e. had no structure, and also had spin-$\frac{1}{2}$. However, their similarity was marred by the fact that the leptons numbered four particles (ν_μ, μ; ν_e, e) but the quarks numbered only three particles (u, d, s): this suggested the existence of a fourth quark.

Since free quarks were known not to exist, physicists attempted to search for evidence of charm flavored quarks within more massive hadrons. Evidence for the fourth quark, termed earlier the *charmed* quark by Glashow and Bjorken, (c), was found in 1974 by two different experimental groups led by Burton Richter (1931-2018) and Samuel Ting (1936-), respectively. This led to the introduction of a new additive quantum number, C, called charm, which was conserved in both the strong nuclear and electromagnetic interactions but like S not in CC weak nuclear interactions. The charmed quark had $\mathbf{I} = 0$, $Q = +\frac{2}{3}$, $A = \frac{1}{3}$, $S = 0$ and $C = +1$, corresponding to a generalization of Eq. (4.10) to

$$Q = I_3 + \frac{1}{2}(A + S + C). \tag{4.11}$$

Subsequently, evidence for additional quarks, bottom (b) and top (t), was found in 1977 and 1995, respectively. This led to the need to introduce yet more additive quantum numbers, bottomness B and topness T, and a further generalization of Eq. (4.11) to

$$Q = I_3 + \frac{1}{2}(A + S + C + B + T). \tag{4.12}$$

The bottom quark had $\mathbf{I} = 0$, $Q = -\frac{1}{3}$, $A = \frac{1}{3}$, $S = 0$, $C = 0$, $B = -1$, $T = 0$, while the top quark had $\mathbf{I} = 0$, $Q = +\frac{2}{3}$, $A = \frac{1}{3}$, $S = 0$, $C = 0$, $B = 0$, $T = +1$, so that Eq. (4.12) was satisfied for both quarks. The six quarks, u, d, c, s, t and b are six *elementary* particles of the current SM [2-4,6,7]. Table 4.1 gives the quantum numbers allotted to the six quarks of the SM.

quark	Q	I	I_3	A	S	C	B	T
u	$+\frac{2}{3}$	$\frac{1}{2}$	$+\frac{1}{2}$	$\frac{1}{3}$	0	0	0	0
d	$-\frac{1}{3}$	$\frac{1}{2}$	$-\frac{1}{2}$	$\frac{1}{3}$	0	0	0	0
c	$+\frac{2}{3}$	0	0	$\frac{1}{3}$	0	+1	0	0
s	$-\frac{1}{3}$	0	0	$\frac{1}{3}$	-1	0	0	0
t	$+\frac{2}{3}$	0	0	$\frac{1}{3}$	0	0	0	+1
b	$-\frac{1}{3}$	0	0	$\frac{1}{3}$	0	0	-1	0

Table 4.1. Quantum numbers for quarks in SM.

Equation (4.12) is the culmination of the strong isospin concept developed within the SM framework, since Heisenberg introduced it in 1932. It implies a very badly broken flavor $SU(6)_f$ symmetry, with conservation of I_3, A, S, C, B and T in both strong and electromagnetic interactions. These conservation laws express the fact that the various kinds (flavors) of quarks are strictly conserved in all except CC weak nuclear interaction processes. Furthermore, each type of quark is subject to the same strong nuclear (color) force. The flavors possessed by a hadron (baryon or meson) are correlated with the number of quarks having those flavors comprising it, In particular, a hadron's strong isospin projection quantum number is determined by the number of up and the number of down quarks in it. The primary symmetry breaking arises from the different quark masses.

4.3 Weak Isospin

As discussed in Chapter 3, the first weak nuclear interaction process, nuclear β-decay, was discovered as early as 1896. In 1930 Pauli had essentially proposed that the continuous energy spectra of the electrons emitted in the radioactive decays of certain nuclei could be understood if a neutron decayed to a proton with the emission of both an electron and another particle, later termed an electron antineutrino:

$$n^0 \to p^+ + e^- + \bar{\nu}_e. \tag{4.13}$$

This raised the question: What is the nature of the force that causes this radioactive decay?

In 1934 Fermi proposed that in β-decay, a neutron decays to a proton in a manner analogous to the emission of a photon in an electromagnetic interaction. Fermi's theory assumed that the decay process took place at a single spacetime point, corresponding to the short-range nature of the underlying weak nuclear force. However, in a radiative transition the photon is the mediating particle of the electromagnetic interaction, but it was difficult to understand how the corresponding electron-antineutrino pair could be the weak nuclear force quantum.

In 1938 Klein suggested that the weak nuclear force could be mediated by massive charged bosons, now called W^+ and W^- bosons, which had properties similar to those of photons. He termed them 'electrically charged photons' but unlike photons, they were *massive* in order to satisfy the very short-range nature of the weak nuclear forces. Thus β-decay could be considered to be a two-step process:

$$n^0 \to p^+ + W^-, \qquad W^- \to e^- + \bar{\nu}_e, \tag{4.14}$$

provided the large mass of the W^- boson and its short lifetime are compatible with Heisenberg's uncertainty principle (see Section 3.3). Such weak nuclear interactions, involving W^+ and W^- mediating bosons, are known as charge-changing (CC) weak nuclear interactions.

Between 1947 and 1953 several new particles were discovered in cosmic rays. In particular, two of these particles then known as the tau particle, which decayed into three pions:

$$\tau^+ \rightarrow \pi^+ + \pi^+ + \pi^-, \tag{4.15}$$

and the theta particle, which decayed into two pions:

$$\theta^+ \rightarrow \pi^+ + \pi^0, \tag{4.16}$$

presented a problem. Both particles decayed via a CC weak nuclear force and were *indistinguishable* apart from their decay mode, since their masses and lifetimes were found to be about the same.

The essential problem was provided by Richard Dalitz (1925-2006), who argued that the τ^+ particle would have parity -1, while the θ^+ particle would have parity $+1$, if the pions had parity -1 as was generally believed at that time. Hence, if conservation of parity holds, the theta having parity $+1$, and the tau having parity -1, could not be the same particle. This was known as the *theta-tau puzzle*.

In 1924 Otto Laporte (1903-1971) discovered that the energy levels of iron atoms consist of two subsets that do not intercombine [6]. These observations were explained in 1927 by Eugene Wigner (1902-1995), who essentially introduced the notion of 'parity' into quantum mechanics. Wigner divided atomic states into 'normal states' (actually states of positive parity) and 'reflected states' (actually states of negative parity) and noted that for the dominant electromagnetic transitions (electric dipole radiation) only transitions between normal and reflected states are allowed, since such transitions involve a change of one unit of orbital angular momentum, i.e. $\Delta l = \pm 1$. Wigner noted that if parity was a conserved property, it required in quantum mechanics the existence of a unitary operator P that commuted with the Hamiltonian, and taking P to be a 'reflection' operator, its eigenvalues can be only ± 1.

The physical quantity *parity* is a purely quantum mechanical concept, since it is related to the properties of wave functions, if:

$$\psi(-x, -y, -z) = \psi(x, y, z), \tag{4.17}$$

where x, y, z are the spatial coordinates and ψ is the value of the wave function at a given point, so that ψ is *symmetric*, then ψ is assigned a positive parity $P = +1$. If however:

$$\psi(-x, -y, -z) = -\psi(x, y, z), \tag{4.18}$$

so that ψ is *antisymmetric*, then ψ is assigned a negative parity $P = -1$. It should be noted that parity is *not* an additive quantum number but is a *multiplicative* quantum number, i.e. one for which the corresponding 'product' rather than the 'sum' of the quantum numbers of a system of particles tends to be conserved.

In 1956 Tsung-Dao Lee (1926-) and Chen-Ning Yang (1922-), in order to resolve the theta-tau puzzle, proposed that parity conservation might be violated in weak nuclear interactions. This was rapidly confirmed in 1957 by several groups, including Chien-Shiung Wu (1912-1997) and collaborators, who studied the β-decay of Co^{60}.

The discovery of parity violation suggested that the CC weak nuclear interaction consisted of two terms with opposite parities. During 1957 it was shown that the so-called 'V-A' theory of the CC weak nuclear interaction [6], developed by George Sudarshan (1931-2018) and Robert Marshak (1916-1992) described the observed parity violations in terms of a vector (V) interaction with negative parity and an axial vector (A) interaction with positive parity. In 1958 Feynman and Gell-Mann published a similar V-A theory of the CC weak nuclear interaction.

In 1936 the muon was discovered by Anderson and Neddermeyer and in 1947 the pion was found independently by both Perkins and by Powell and his group. Both these particles decay via CC weak nuclear interaction processes, such as:

$$\pi^- \to \mu^- + \bar{\nu}_\mu \tag{4.19}$$

and

$$\mu^- \to e^- + \bar{\nu}_e + \nu_\mu . \tag{4.20}$$

On the other hand, certain decay modes were *not* observed. In particular, the muon decay modes:

$$\mu^- \to e^- + \gamma \tag{4.21}$$

and

$$\mu^- \to e^- + e^+ + e^- . \tag{4.22}$$

In order to explain the absence of such decay modes, Emil Konopinski (1911-1990) and Hormoz Mahmoud (1918-2010) in 1953 introduced the idea of *lepton conservation* analogous to baryon conservation. By assigning lepton number $L = +1$ to e^-, μ^+ and ν (the different kinds of neutrinos were not yet established), $L = -1$ to e^+, μ^- and $\bar{\nu}$ and $L = 0$ to all other particles, the processes (4.21) and (4.22) were forbidden. In addition, conservation of lepton number implied that the two neutral particles emitted in muon decay had the *same* lepton number, later shown to be false. Lepton number was found to be conserved in the reaction

$$\bar{\nu}_e + p^+ \to n^0 + e^+ \,, \tag{4.23}$$

which provided evidence for the existence of the electron antineutrino. The reaction

$$\bar{\nu}_e + n^0 \to p^+ + e^- \,, \tag{4.24}$$

violating lepton number conservation, was not observed.

In 1962 Leon Lederman (1922-2018), Melvin Schwartz (1932-2006) and Jack Steinberger (1921-2020) demonstrated that the neutrinos (ν_μ) produced in the decay of π^+ mesons:

$$\pi^+ \to \mu^+ + \nu_\mu \,, \tag{4.25}$$

subsequently gave rise to the production of muons via the reaction

$$\nu_\mu + n^0 \to p^+ + \mu^- \,. \tag{4.26}$$

However, the corresponding reaction in which electrons are produced was not observed, i.e. there are no reactions of the form

$$\nu_\mu + n^0 \to p^+ + e^- \,. \tag{4.27}$$

In the scheme of Konopinski and Mahmoud, the ν_μ produced in the decay of a π^+ meson would be interpreted as an antineutrino $(\bar{\nu})$ in order to conserve lepton number. This is consistent with reaction (4.26) being allowed and reaction (4.27) being forbidden. However, in the 1962 experiment *both* π^+ and π^- mesons were present. The latter produces neutral particles with opposite lepton number to ν_μ, so that both μ^- and e^- could be expected to be produced in the experiment. The results showed that the neutrino associated with the muon was *different* from the neutrino associated with the electron.

The above experimental evidence for the existence of two neutrinos led to the acceptance of an alternative scheme, involving separate lepton numbers, $L_e = +1$ and $L_\mu = 0$ for the lepton pair (ν_e, e^-) and $L_e = 0$ and $L_\mu = +1$ for the lepton pair (ν_μ, μ^-), respectively. If these additive quantum numbers are assumed to be separately conserved in all interactions, then all the unobserved decay modes mentioned above are forbidden. This latter scheme, unlike the earlier one, is readily extended to the lepton pair (ν_τ, τ^-), involving a third charged lepton, the tau particle (τ^-), discovered by Martin Perl (1927-2014) and collaborators in 1975 and an associated neutrino, the tau neutrino (ν_τ), discovered by the same group in 2000.

Table 4.2 gives the allotted additive quantum numbers in the SM for the six leptons. Here

$$L = L_e + L_\mu + L_\tau, \tag{4.28}$$

is the total lepton number, which is also conserved in all CC weak nuclear interaction processes. It should be noted that the total lepton number, L, differs from that introduced earlier by Konopinski and Mahmoud. The six leptons, ν_e, e^-, ν_μ, μ^-, ν_τ and τ^- are six *elementary* particles of the current SM.

lepton	Q	L	L_e	L_μ	L_τ
ν_e	0	1	1	0	0
e^-	-1	1	1	0	0
ν_μ	0	1	0	1	0
μ^-	-1	1	0	1	0
ν_τ	0	1	0	0	1
τ^-	-1	1	0	0	1

Table 4.2. Quantum numbers for leptons in SM.

To summarize: the introduction of lepton numbers, which were strictly conserved in both electromagnetic and CC weak nuclear interactions, provided a very useful description of the allowed decay modes and the possible reactions involving leptons.

Another property of the weak nuclear forces discovered in the late 1940's was their 'universality'. Analysis of experiments revealed that the coupling constants (i.e. the strengths of the forces) for μ-decay and μ-capture were of the same order of magnitude as that for β-decay. This led to the hypothesis of a *universal weak nuclear force* [6], mediated by the W bosons.

The occurrence of the three doublets (ν_e, e^-), (ν_μ, μ^-) and (ν_τ, τ^-) with separate lepton numbers (see Table 4.2) and their similar behavior with respect to the 'universal' CC weak nuclear force mediated by the W^+ and W^- particles led naturally to the notion of a *weak isospin*, associated with an $SU(2)$ symmetry.

The concept of weak isospin associated with the CC weak nuclear force developed considerably later than that of strong isospin associated with the strong nuclear force. This was mainly because the weak isospin symmetry in nature is more obscure than the corresponding strong isospin symmetry, although it is more widespread, in the sense that both leptons and quarks are subject to the CC weak nuclear interaction.

In 1957 Schwinger, following the ideas of Klein, suggested a triplet of vector fields (W^+, γ, W^-), whose universal couplings generated both the CC weak nuclear force, mediated by the charged massive vector bosons, W^+ and W^-, and the electromagnetic force, mediated by the photon, γ. This suggestion was based upon the notion that the CC weak nuclear force was a *fundamental* force like the electromagnetic force and that this force arose from an $SU(2)$ local gauge theory (see Chapter 5). This endeavor by Schwinger was the first attempt to unify the electromagnetic and the CC weak nuclear force. However, it suffered from the fact that the large masses of the W bosons, required to account for the very short-range nature of the CC weak nuclear force, had to be inserted into the theory 'by hand' in conflict with the gauge invariance requirement of the theory that the mediating particles should be *massless*.

In 1958 Sidney Bludman (1927-) proposed that many aspects of the weak nuclear force could be described by an $SU(2)$ global gauge theory assuming a triplet of three vector bosons, W^+, W^0 and W^-, in a 'weak isospin space' (see Section 5.2). Moreover, it did indicate the possibility of a *neutral* weak nuclear force, mediated by the W^0 boson, which is distinct from the usual electromagnetic force.

Bludman also noted that, if the lepton doublets (ν_e, e^-) and (ν_μ, μ^-) were considered as weak isospin doublets $(\mathbf{i} = \frac{1}{2})$ with ν_e and ν_μ having $i_3 = +\frac{1}{2}$ and e^- and μ^- having $i_3 = -\frac{1}{2}$, the charge of each lepton satisfied a relation analogous to that for strong isospin [Eq. (4.1)]:

$$Q = i_3 - \frac{1}{2}L, \tag{4.29}$$

where $L = +1$ is the lepton number for each lepton, analogous to $A = +1$ for each baryon. Indeed the occurrence of the three doublets (ν_e, e^-), (ν_μ, μ^-) and (ν_τ, τ^-) with separate lepton numbers (see Table 4.2) and their similar behavior with respect to the 'universal' CC weak nuclear force mediated by the W bosons naturally suggests the notion of a weak isospin with an $SU(2)$ symmetry.

4.4 Cosmology

During the modern era considerable progress was made in understanding the nature of the Universe, employing primarily both the general theory of relativity to describe the physics of the very large, dominated by the gravitational force, and quantum mechanics to describe nuclear reactions and what are the basic building blocks of the Universe.

As discussed briefly in Section 3.2, in 1922 Friedmann, assuming that the distribution of ordinary matter in the Universe was homogeneous and isotropic, showed that the Universe was either expanding or contracting according to the equations of general relativity.

In 1927 Lemaître obtained similar results to Friedmann but noted that expanding universes could be extrapolated backwards in time to an originating singular point that became associated with the so-called 'Big Bang'. Lemaître called the originating singular point the 'primordial atom', which he considered to be an extremely small and compact system of very dense and very hot matter. This model of the origin of the Universe was termed the Big Bang model by Fred Hoyle (1915-2001) in 1950.

As discussed also in Section 3.2, in 1929 Hubble discovered that light from galaxies was redshifted, implying that these galaxies were receding from the Earth, if one assumed that the redshift was due to a Doppler effect. Hubble observed that there was a linear relationship between the radial speed with which a galaxy recedes from the Earth and its distance from the Earth. If the Universe is expanding, this implies that (i) only expanding solutions of the Friedmann equations are allowed as solutions for the Universe and (ii) the Universe must have had a very dense and hot beginning. These observations supported Lemaître's ideas concerning the origin of the Universe. It should be noted that the expanding solutions of Friedmann consider that it is space itself that is expanding and that the galaxies are at rest within the expanding space. Thus the redshift for each galaxy is a consequence of the wavelength of the light being stretched by the expansion of space and is *not* a normal Doppler redshift.

The general theory of relativity describes spacetime by a metric that determines the distances separating nearby points (stars, galaxies, etc). The assumption that the metric should be homogeneous and isotropic on large scales uniquely requires that the metric be the Friedmann-Lemaître-Robertson-Walker (FLRW) metric. During 1935-1937 Howard Robertson (1903-1961) and Arthur Walker (1909-2001) proved that the FLRW metric is the only one that is spacially homogeneous and isotropic.

The prevailing model of the Big Bang is based upon general relativity. According to this theory, extrapolation of the expansion of the Universe backwards in time yields an infinite mass-energy density and temperature at a finite time, approximately 13.8 billion years ago. This was demonstrated by Roger Penrose (1931-) and Stephen Hawking (1942-2018) in 1970 [18]. Thus the 'birth' of the Universe appears to be associated with a *singularity*, which describes not only a breakdown of the theory of general relativity, but also all the laws of physics. This suggests that the theory of general relativity with the FLRW metric is *not valid* for extremely small regions of space.

On the other hand, the Big Bang scenario has had some success. In 1948, George Gamow (1904-1968) suggested that the present features of the universe could be understood as a result of the evolutionary development of the universe via expansion from the Big Bang phase. In particular, he suggested that the elements could have been made during the early hot matter-energy phase associated with the Big Bang. It has since been shown that as the initial hot dense mass-energy phase of the universe cooled during the expansion that only several light elements were formed, including hydrogen ($\approx 75\%$), helium ($\approx 25\%$) and small amounts of deuterium, lithium, etc., of the mass of the Universe.

As the hot dense phase continued to cool down during the expansion, the atomic nuclei of hydrogen, helium, etc., captured electrons thereby generating *neutral* atoms. This is estimated to have occurred about 400,000 years after the Big Bang, when photons ceased interacting significantly with matter, leading to the occurrence of the so-called *Cosmic Microwave Background* (CMB) radiation. In 1948 Ralph Alpher (1921-2007) and Robert Herman (1914-1997) calculated the present temperature of this CMB to be about 5 K, remarkably close to the modern value of about 2.73 K, determined by the COBE satellite. In addition, the COBE results showed an extremely isotropic and homogeneous CMB.

In the 1930s, astrophysicists were more interested in how the Sun and stars shone, rather than in the origin of the Universe. It was hoped that the new theories of relativity and quantum mechanics would provide an understanding of how stars emitted energy.

In 1925 Cecilia Payne (1900-1979) found from her studies of stellar spectra that stars were composed mainly of hydrogen and helium, but predominantly of hydrogen. This was contrary to the prevailing theory at that time, which considered that stars like the Sun consisted of the same elemental composition as the Earth, so that Payne's conclusion was disputed.

However, other astronomers gradually recognized that Payne was essentially correct, finding that the Sun's mass is composed of about 70% hydrogen, 28% helium and 2% heavier elements.

Eddington and his colleagues suggested that the Sun converted four hydrogen nuclei into one helium nucleus, thereby producing energy from the atomic mass lost in this process, according to $E = mc^2$. They estimated that this process could proceed steadily for some 10 billion years.

In 1938 Hans Bethe (1906-2005) showed how four protons could be converted into a helium nucleus within the Sun via a chain of nuclear reactions involving the presence of a C^{12} nucleus. Later Bethe and Charles Critchfield (1910-1994) indicated an alternative route for the above process: a chain of nuclear fusion reactions not involving C^{12}, which is the dominant process for powering stars like the Sun:

$$H^1 + H^1 \rightarrow H^2 + e^+ + \nu_e, \qquad (4.30)$$
$$H^2 + H^1 \rightarrow He^3 + \gamma,$$
$$He^3 + He^3 \rightarrow He^4 + 2H^1.$$

In 1948 Herman Bondi (1919-2005), Thomas Gold (1920-2004) and Hoyle suggested an alternative theory to the Big Bang scenario. This was called the 'steady state theory', which considered that the Universe was essentially eternal, in the sense that it looked roughly the same at all times as well as at all points of space. The theory assumed that the galaxies were receding from the Earth as concluded by Hubble from the observed redshifts and that the gaps in between were replaced by new galaxies continually forming from new matter that was continually being created.

In 1964 two radio astronomers, Arno Penzias (1933-) and Robert Wilson (1936-) accidently discovered the CMB. This discovery caused the majority of cosmologists to accept the Big Bang scenario, rather than steady-state type theories, since Alpher and Herman had predicted that there should be a detectable echo (the CMB) of the Big Bang in 1948.

This acceptance of the Big Bang scenario took place in spite of several perceived problems with it. First, the Doppler interpretation of the cosmological redshift may be incorrect. In 1937 Hubble had doubts about the Doppler interpretation of the redshifts, since he considered that this led to strange and dubious conclusions. On the other hand, Hubble considered that the 'tired light' interpretation of the redshifts, as proposed by Zwicky in 1929 (see Section 3.2), led to conclusions that seemed more plausible. The tired light theory considers that the CMB radiation is simply tired light in the microwave band, and this theory has recently been revived [21].

Second, all three cosmological models that we have discussed so far, the Big Bang, the steady state theory of Hoyle *et al.* and Zwicky's tired light theory, all suffer from one basic problem: How did the mass-energy of the Universe originate in the first place?

The current SMC assumes that the Universe was created in the Big Bang from pure energy, and is now composed of about 5% ordinary matter, 27% dark matter and 68% dark energy [1].

While the SMC is based primarily upon two theoretical models: (1) the SM of particle physics, which describes the physics of the very small in terms of quantum mechanics (see Chapter 5) and (2) the general theory of relativity, which conventionally describes the physics of the very large, dominated by the gravitational force, in terms of classical mechanics (see Section 3.2), it also depends upon two major dubious assumptions: the existence of both *dark matter* and *dark energy*, which the SMC claims constitute about 27% and 68%, respectively of the mass-energy content of the Universe.

During the era of modern physics, several astronomical observations concerning the motions of galaxies in a cluster of galaxies, the structure of galaxies and the rotation of stars within spiral galaxies, could not be understood in terms of Newton's universal law of gravitation and the visible atomic (ordinary) matter within the galactic systems.

It was concluded that such observations could only be described satisfactorily if there existed stronger gravitational fields than those provided by the visible matter and Newtonian gravity. Such gravitational fields required either more mass or an appropriate modification of Newton's universal law of gravitation.

Early evidence for such a 'mass discrepancy' was observed in 1933 by Zwicky for the Coma cluster of galaxies. Zwicky introduced the term *dark matter* for the mass discrepancy and estimated that the cluster contained considerably more dark matter than the visible galactic matter in order to account for the fast motions of the galaxies within the cluster and also to hold the cluster together.

In the 1970s Vera Rubin (1928-2016), Kent Ford (1931-) and Norbert Thonnard (1943-2014) measured the 'rotation curve' initially for the Andromeda galaxy, the nearest spiral galaxy to the Milky Way, and later many other spiral galaxies. The rotation curve of a galaxy is the dependence of the orbital velocity of the visible matter in the galaxy on its radial distance r from the center of the galaxy. The observations found that the measured rotation curves disagreed with those expected from Newton's gravitational law for large r: instead of falling off as $1/\sqrt{r}$ as occurs for the planets in the Solar System, as predicted by Newton's law, the measured rotation curves of the observed galaxies were essentially 'flat' at large distances from the center of each galaxy. In 1970, Kenneth Freeman (1940-) had also found rotation curves for several galaxies that disagreed with expectation based upon the assumption that the galaxies consisted of stars, hydrogen gas, and nothing else. Freeman suggested that these galaxies, like the Coma cluster observed much earlier by Zwicky contained considerably more invisible 'dark matter' than luminous matter. In addition several observers, using radio telescopes, measured the 21cm rotation curves of the neutral hydrogen gas that extended far beyond the luminous matter of each galaxy. In all cases, the complete rotation curve was essentially flat out to the edge of the 21cm data.

This led to the introduction of the "dark matter hypothesis" by Jeremiah Ostriker (1937-), James Peebles (1935-) and Amos Yahil (1943-), who concluded that the rotation curves of spiral galaxies could most plausibly be understood if the spiral galaxy was embedded in a giant spherical halo of invisible 'dark matter' that provided a large contribution to the gravitational field at large distances from the center of the galaxy. Ostriker and Peebles had found earlier that their computer models of a typical spiral galaxy indicated that such a galaxy fell apart, if the gravitational field arose solely from the luminous matter in the galaxy: they concluded that the stability of a typical spiral galaxy required more dark matter to provide more gravity.

This dark matter hypothesis is very dubious, since to date no dark matter has been definitely *detected* and the nature of dark matter remains *unknown*. This will be discussed in more detail in Chapter 8.

The notion of 'dark energy' arose from observations carried out by two independent teams of astronomers that suggested that the expansion of the universe is *accelerating*. One team led jointly by Adam Riess (1969-) and Brian Schmidt (1967-), and the other team led by Saul Perlmutter (1959-) published their results in 1998 and 1999, respectively. These observations were very surprising and unexpected, since it was generally considered that the spatial expansion of the universe should be slowing down due to the gravitational attraction of the galaxies.

Both sets of observations were based upon the analysis of supernovae of Type Ia, which are considered to be excellent standard candles across cosmological distances and allow the expansion history of the universe to be measured by considering the relationship between the distance to an object and its redshift, which indicates how fast the supernova is receding from us. Both teams found that the supernovae observed about halfway across the observable universe (6-7 billion light-years away) were dimmer than expected and concluded that the expansion of the universe was accelerating rather than slowing down as expected.

The conclusion from this observation was that the universe had to contain enough energy to overcome gravity. This energy was named 'dark energy', which is considered by some astrophysicists to be a hypothetical form of energy that pervades the whole of space and causes the expansion of the Universe to accelerate at large cosmological distances.

Currently there exists no accepted physical theory of dark energy, suggesting that the existence of such energy is a dubious assumption of the SMC. This will be discussed in more detail in Chapter 8.

Chapter 5

Standard Model of Particle Physics

5.1 Introduction

In this chapter, the current formulation of the Standard Model of particle physics (SM) [2–4, 6, 7], which was essentially finalized in the mid-1970s following the experimental confirmation of quarks, will be discussed critically. In particular, the *incompleteness* of the SM, associated with several dubious assumptions, will be considered in order to progress beyond the SM and to replace it with an alternative model, the Generation Model (GM).

However, first I shall briefly summarize the main historical developments of the SM, involving the various building blocks, the leptons and quarks, as well as the various forces found to occur between these building blocks, as discussed in Chapters 2-4.

In the era of classical physics (prior to 1895), the elementary particles were atoms and the only known fundamental forces were the gravitational force and the electromagnetic force. The nature of the electromagnetic force changed significantly from the earlier idea of objects exerting a force upon each other to the notion of a field containing energy acting between two objects.

In the era of transitional physics (1895-1932), significant progress was made in the understanding of the nature of both matter and forces by the development of two new theories, relativity theory and quantum theory, and also the development of new apparatus, including the Crookes tube and the Wilson cloud chamber.

In particular, the atom was found to consist of three kinds of subatomic particles, electrons, protons and neutrons, arranged so that each atom is electrically neutral, consisting of a minute nucleus, composed of Z protons and *A-Z* neutrons, surrounded by a cloud of *Z* electrons. Thus, the proton, the neutron and the electron became the elementary particles of matter.

In addition, the special theory of relativity replaced the Galilean relativity of Newtonian mechanics, leading to consequences associated with the nature of forces, including the variation of mass with velocity and the equivalence of mass and energy. The relativistic variation of mass leads to an advance of the perihelion of an elliptical orbit associated with an inverse square force law such as Newton's universal law of gravity. Indeed, the anomalous precession of the perihelion of Mercury was described by Einstein's general theory of relativity, which incorporated the special theory of relativity, thus inspiring confidence in Einstein's new theory of gravity.

The introduction of the quantum hypothesis by Planck in 1900 initiated the development of a quantum theory that provided a basis for understanding the subatomic world. In particular, in 1905 Einstein proposed that electromagnetic radiation consisted of energy quanta, later called photons. This implied that the electromagnetic field, suggested by Faraday, consisted of photons, so that the electromagnetic force became viewed as the exchange of photons.

In 1928 Dirac provided a unification of quantum mechanics and special relativity by deriving a relativistic wave equation called the Dirac equation that indicated that matter consists of both particles and antiparticles. Furthermore, the observation of phenomena that disagreed with the predictions of the Dirac equation led to the development of a relativistic quantum field theory called QED, which describes the interaction of photons with electrons and positrons.

In the era of modern physics, especially following the development of powerful accelerators, such as the Brookhaven Cosmotron in 1952, many new particles, which were subject to the strong nuclear force, called hadrons, were created. Following the discovery of so many hadrons. considerable effort was made to understand and classify the various kinds of new particles, resulting in new models, including the *eightfold way* suggested by Gell-Mann and Ne'eman in 1961, which led to the *quark model* proposed independently by Gell-Mann and Zweig in 1964. The quark model considered that all the hadrons known at that time were composed of three elementary particles, called quarks by Gell-Mann: the up (u), down (d) and strange (s) quarks, together with their corresponding antiparticles: $\bar{u}$, $\bar{d}$ and $\bar{s}$. Each meson was composed of a quark-antiquark pair, while each baryon was composed of three quarks and each antibaryon was composed of three antiquarks. The quark model was based upon what became known as an $SU(3)_f$ flavor symmetry. Following the discovery of evidence for the existence of three more quarks, the charmed quark (c) in 1974, the bottom quark (b) in 1977 and the top quark (t) in 1995, the model was extended to an $SU(6)_f$ flavor symmetry, and the whole set of six quarks formed six elementary particles of the SM.

Unfortunately, the 1964 quark model indicated that several hadrons were predicted to have three identical particles in the same quantum state, thereby violating the Pauli exclusion principle. This problem was resolved in 1965 by Han and Nambu, who introduced a new degree of freedom for each quark called color charge that could take three values, red, green or blue, while the antiquarks carried anticolor charges, antired, antigreen or antiblue. The quarks and antiquarks were assumed to interact via an octet of gluons. The gluon fields surrounding an isolated color charge were considered to be such that an isolated quark would have an infinite energy associated with it. This was the reason why isolated quarks are not seen. In hadrons, isolated quarks were avoided in two ways: either by bringing an antiquark close to a quark, thus forming a colorless meson, or by bringing three quarks, one of each color, together, forming a baryon, so that in each case the composite hadron is colorless. In 1973 the chromodynamic force between particles carrying a color charge was developed into a more complete theory, called QCD, by Fritzsch, Gell-Mann and Leutwyler. In QCD, the chromodynamic force is mediated by eight electrically neutral massless spin-1 gluons. Each gluon carries both a single color charge and a single anticolor charge, and consequently gluons exert chromodynamic forces upon each other. These self-interactions of the gluons lead to two important consequences: (1) *antiscreening effects* leading to an increase in the strong nuclear force field as the separation between the quarks increases and (2) *color confinement* leading to a finite range of the strong nuclear force.

In the modern era, the nature of the CC weak nuclear force that causes the radioactive β-decay of a neutron was investigated. This force is considered to be mediated by a massive charged boson, W^-, following a suggestion by Klein in 1938.

In 1956 the theta-tau puzzle, associated with the decay of a particle then known as the tau particle that decayed into three pions, implying negative parity of the tau particle and the decay of a particle then known as the theta particle that decayed into two pions, implying positive parity of the theta particle, led to the nonconservation of parity in CC weak nuclear interactions. Both the theta and tau particles decaying via a weak nuclear force were indistinguishable apart from their decay mode, since their charges, masses and lifetimes were found to be the same. If the CC weak nuclear force conserved parity then the theta and tau particles could not be the same particle. In 1956 Lee and Yang proposed that parity conservation may be violated in weak nuclear force decays. This was confirmed in 1957 by several groups and during 1957 Sudarshan and Marshak showed that the observed parity violations could be described in terms of a V-A interaction.

During the modern era, several pairs of particles, similar to the electron (e^-) and the electron neutrino (ν_e) were discovered: the muon (μ^-) and the muon neutrino (ν_μ) were found in 1962, while the tau (τ^-) and the tau neutrino (ν_τ) were discovered in 1975 and 2000, respectively. These six particles are called leptons, from the Greek $\lambda\epsilon\pi\tau o\sigma$ meaning 'small' and form six additional elementary particles of the SM. These leptons are classified in terms of lepton numbers, which are conserved in all interactions of the SM.

In the modern era, attempts were made to describe the universal CC weak nuclear force in terms of a weak isospin, analogous to the strong isospin employed for the strong nuclear force. The concept of weak isospin was successfully used to incorporate the lepton number additive quantum numbers in the classification of the leptons within the SM. It also provided the basis for the proposal to unify the electromagnetic force with the CC weak nuclear force (see Section 5.2).

5.2   Elementary Particles and Fundamental Forces of SM

In the SM the elementary particles that are the constituents of ordinary matter are assumed to be the six leptons: electron neutrino (ν_e), electron (e^-), muon neutrino (ν_μ), muon (μ^-), tau neutrino (ν_τ), tau (τ^-) and the six quarks: up (u), down (d), charmed (c), strange (s), top (t) and bottom (b), together with their antiparticles. These twelve elementary particles are all spin-$\frac{1}{2}$ particles and fall naturally into three families or generations: (i) ν_e, e^-, u, d ; (ii) ν_μ, μ^-, c, s ; (iii) ν_τ, τ^-, t, b . Each generation consists of two leptons with charges $Q = 0$ and $Q = -1$ and two quarks with charges $Q = +\frac{2}{3}$ and $Q = -\frac{1}{3}$. The masses of the particles increase significantly with each generation with the possible exception of the neutrinos, whose very small masses have yet to be determined.

In the SM the leptons and quarks are allotted several additive quantum numbers as discussed in Sections 4.2 and 4.3. These are given in Table 5.1. Again, for each particle additive quantum number N, the corresponding antiparticle has the additive quantum number $-N$.

Particle	Q	L	L_e	L_μ	L_τ	A	S	C	B	T
ν_e	0	1	1	0	0	0	0	0	0	0
e^-	-1	1	1	0	0	0	0	0	0	0
ν_μ	0	1	0	1	0	0	0	0	0	0
μ^-	-1	1	0	1	0	0	0	0	0	0
ν_τ	0	1	0	0	1	0	0	0	0	0
τ^-	-1	1	0	0	1	0	0	0	0	0
u	$+\frac{2}{3}$	0	0	0	0	$\frac{1}{3}$	0	0	0	0
d	$-\frac{1}{3}$	0	0	0	0	$\frac{1}{3}$	0	0	0	0
c	$+\frac{2}{3}$	0	0	0	0	$\frac{1}{3}$	0	1	0	0
s	$-\frac{1}{3}$	0	0	0	0	$\frac{1}{3}$	-1	0	0	0
t	$+\frac{2}{3}$	0	0	0	0	$\frac{1}{3}$	0	0	0	1
b	$-\frac{1}{3}$	0	0	0	0	$\frac{1}{3}$	0	0	-1	0

Table 5.1. SM additive quantum numbers for leptons and quarks.

Table 5.1 shows that, except for charge, Q, leptons and quarks are allotted *different* kinds of additive quantum numbers so that this classification of the elementary particles of the SM is *nonunified*.

The additive quantum numbers Q and A are assumed to be conserved in the electromagnetic and both the strong and CC weak nuclear forces. The lepton numbers L, L_e, L_μ and L_τ are not involved in the strong nuclear force but are strictly conserved in both the electromagnetic and weak nuclear forces. The remainder, S, C, B and T are strictly conserved only in the strong nuclear and electromagnetic forces but may undergo a change of one unit in the weak nuclear force.

The introduction of a 'partially conserved' additive quantum number such as strangeness in the development of the SM was a very dubious assumption. In quantum mechanics, quantum numbers are usually conserved quantities and the nature of the CC weak nuclear interaction is 'weak' because it is mediated by massive W bosons not because the strangeness quantum number is not conserved. The strangeness assumption led to several problems for the development of the SM, associated with both the classification of the elementary particles and the nature of the universality of the CC weak nuclear force.

The SM recognizes four fundamental forces in nature: the strong nuclear, electromagnetic, CC weak nuclear, and gravitational forces. Since the gravitational force plays no role in particle physics, because it is so much weaker than the other three fundamental interactions, the SM does not attempt to explain the gravitational force. In the SM the other three fundamental forces are assumed to be associated with a local gauge field.

In elementary particle physics, symmetry and gauge invariance play a major role in the understanding of the elementary particles and the forces between them. I shall now give a brief outline of the physics involved in the various symmetries and gauge invariances assumed in the SM, avoiding the rather complex mathematical details describing these concepts.

The concept of gauge invariance as a physical principle governing the fundamental interactions between elementary particles was first proposed in 1919 by Hermann Weyl (1885-1955) in an attempt to extend ideas employed by Einstein's general theory of relativity, involving the gravitational force, to the case of the electromagnetic interaction. Indeed, Einstein spent the latter part of his life trying to incorporate electromagnetism into his general theory of relativity by describing electric and magnetic fields as properties of spacetime [3]. This attempt by Einstein to unify the gravitational and electromagnetic forces, which had similar inverse square $(1/r^2)$ fields, failed completely as the nature of forces changed with the development of quantum mechanics.

Furthermore, the above initial attempt by Weyl, involving a 'scale invariance' of spacetime, also failed. However, with the development of quantum mechanics, it was realized in 1927 by Vladimir Fock (1898-1974) and Fritz London (1900-1954) that Weyl's original gauge theory could be given a new interpretation: a gauge transformation corresponds to a change in the phase of the wavefunction describing a particle, rather than a change of scale.

There are two kinds of symmetry arising from gauge invariance [22], depending whether the invariance is 'global' or 'local'.

Global gauge invariance leads to a symmetry involving different particles that behave similarly with respect to a particular force. Such a symmetry is called a *flavor* symmetry because the different particles involved are distinguished by some attribute called flavor, which is conserved by the particular force. An example of this kind of symmetry is the flavor $SU(3)_f$ symmetry underlying the eightfold way of Gell-Mann and Ne'eman [19], involving the three quarks: up, down and strange, and the strong nuclear force (see Section 4.2).

Local gauge invariance leads, not only to the conservation of some attribute of a set of particles involved in the symmetry, but also to a *fundamental* force acting between the particles. This force is normally mediated by *massless* particles. An example of this kind of symmetry is the electromagnetic force, associated with a local $U(1)$ gauge invariance, in which electric charge is conserved and the force between charged particles is mediated by massless photons.

In the SM the fundamental strong nuclear force, mediated by massless neutral spin-1 gluons between quarks (antiquarks) carrying a color (anticolor) charge, are described by an $SU(3)$ local gauge theory called QCD, as discussed in Section 4.2. There are eight independent kinds of gluons, each of which carries a combination of a color charge and an anticolor charge. The strong nuclear force between the color (anticolor) charges carried by the quarks (antiquarks) is such that in nature the quarks and antiquarks are grouped into composites of either three quarks of different colors or three antiquarks of different anticolors, called baryons and antibaryons, respectively, or as a composite of a single quark and a single antiquark, called a meson, in which the quark and antiquark carry opposite colors, e.g. red and antired. In the QCD theory, each baryon, antibaryon or meson is *colorless*, i.e. overall has no color or anticolor charge. However, as indicated in Section 4.2, these colorless hadrons may interact strongly via *residual* strong forces arising from their composition of colored quarks and/or antiquarks. On the other hand, the colorless leptons are assumed to be structureless in the SM and consequently do not participate in strong nuclear interactions.

In the SM the fundamental electromagnetic force, mediated by massless neutral spin-1 photons between electrically charged particles, is described by a $U(1)$ local gauge theory called QED (see Section 3.3).

In the SM both the fundamental electromagnetic force and the fundamental strong nuclear force are *universal* in the sense that they act between electrically charged particles and color charged particles, respectively. Initially it was observed that the coupling strength of the CC weak nuclear force for neutron β-decay is about the same as that for μ-decay and for μ-capture. These observations led to the notion of a 'universal' CC weak nuclear force for all CC weak nuclear interaction processes, so that all these three forces possessed a *universal* character.

Subsequently it was found that the CC weak nuclear force involved in strangeness-conserving $(\Delta S = 0)$ decays such as neutron β-decay is in fact slightly weaker than for μ-decay. In addition, the CC weak nuclear force involved in strangeness-changing $(\Delta S = 1)$ decays such as the Λ hyperon β-decay:

$$\Lambda^0 \rightarrow p^+ + e^- + \bar{\nu}_e, \tag{5.1}$$

is weaker still by a significant factor.

In order to preserve the universality of the CC weak nuclear force for both leptonic and hadronic processes, Nicola Cabibbo (1935-2010) assumed that in hadronic processes the CC weak nuclear force is shared between $\Delta S = 0$ and $\Delta S = 1$ transition amplitudes in the ratio of $\cos \theta_c : \sin \theta_c$. This Cabibbo angle θ_c has a value $\approx 13°$. In this way, Cabibbo was able to account for both neutron β-decay and the strangeness-changing decays such as Λ hyperon β-decay employing a universal CC weak nuclear force, since the transition probabilities of these decays are reduced by factors of $\cos^2 \theta_c \approx 0.95$ and $\sin^2 \theta_c \approx 0.05$, respectively, in agreement with experiment.

In the SM, CC weak nuclear force processes involved in pure leptonic decays are governed by the conservation of lepton numbers (see Table 4.2). For example, the decay of a muon:

$$\mu^- \rightarrow e^- + \bar{\nu}_e + \nu_\mu , \tag{5.2}$$

is regarded as a two-step process via the V-A weak interaction involving the W^-:

$$\mu^- \rightarrow \nu_\mu + W^- , \; W^- \rightarrow e^- + \bar{\nu}_e . \tag{5.3}$$

This process is governed by the conservation of muon lepton number (L_μ) in the primary transition and by electron lepton number (L_e) in the secondary transition.

It is convenient to describe these processes by transition amplitudes $a(\nu_\mu, \mu^-; W^-)$ and $a(\nu_e, e^-; W^-)$, respectively. Here $a(x, y; W^-)$ represents the CC weak nuclear force transition amplitude involving the fermions x and y and the W^- boson.

Experiment indicates that the interactions of the electron and its associated neutrino are identical with those of the muon and its associated neutrino. This suggests the notion of the universality of lepton interactions. This concept of a universal CC weak nuclear interaction allows one to write:

$$a(\nu_e, e^-; W^-) = a(\nu_\mu, \mu^-; W^-) = a_w , \tag{5.4}$$

where a_w is the universal CC weak nuclear force transition amplitude. Conservation of L_e and L_μ gives:

$$a(\nu_e, \mu^-; W^-) = a(\nu_\mu, e^-; W^-) = 0 , \tag{5.5}$$

so that there is no CC weak nuclear force between generations in agreement with experiment. The decays involving the tau (τ^-) particle proceed analogously, with the conservation of tau lepton number (L_τ).

In the SM, CC weak nuclear interaction processes involved in both semileptonic and nonleptonic decays are *not* governed solely by the conservation of lepton numbers, since quarks also participate in the decay processes. Table 5.1 lists the additive quantum numbers associated with the six quarks.

Unlike the pure leptonic decays, which are determined by the conservation of the various lepton numbers, there is no additive quantum number in the SM which restricts quark (hadronic) CC weak nuclear forces between generations. In the SM the quarks do not appear to form weak isospin doublets: the known decay processes of neutron and Λ hyperon β-decays suggest that quarks mix between generations and that the 'flavor' additive quantum numbers, S, C, B and T are not necessarily conserved in CC weak nuclear force processes.

In the SM, the universality of the CC weak nuclear force for both leptonic and hadronic processes is restored by adopting the proposal of Cabibbo, discussed earlier, that in hadronic processes the CC weak nuclear force is *shared* between $\Delta S = 0$ and $\Delta S = 1$ transition amplitudes in the ratio of $\cos \theta_c : \sin \theta_c$.

This 'Cabibbo mixing' is an integral part of the SM. In the quark model it leads to a *sharing* of the CC weak nuclear force between quarks of different generations, unlike the corresponding case of leptonic processes. In order to simplify matters, the following discussion will be restricted to the first two generations of the elementary particles of the SM, involving only the Cabibbo mixing, although the extension to three generations, first proposed by Makato Kobayashi (1944-) and Toshihide Maskawa (1940-) in connection with the matter-antimatter asymmetry problem of the universe (see Chapter 9) and indeed prior to the discovery of the third generation, is straightforward. In the latter case, the quark mixing parameters correspond to the Cabibbo-Kobayashi-Maskawa (CKM) matrix elements [3, 7], which indicate that inclusion of the third generation would have a minimal effect on the overall coupling strength of the CC weak nuclear force.

Cabibbo mixing was incorporated into the quark model of hadrons by postulating that the so-called 'weak nuclear force eigenstate quarks', d' and s', form CC weak nuclear force isospin doublets with the u and c quarks, respectively: (u, d') and (c, s'). These weak eigenstate quarks are linear superpositions of the so-called 'mass eigenstate quarks', d and s:

$$d' = d \cos \theta_c + s \sin \theta_c \tag{5.6}$$

and

$$s' = -d \sin \theta_c + s \cos \theta_c . \tag{5.7}$$

The quarks d and s are the quarks which participate in the electromagnetic and the strong nuclear forces with the full allotted strengths of electric charge and color charge, respectively. The quarks d' and s' are the quarks which interact with the u and c quarks, respectively, with the full strength of the CC weak nuclear force.

In terms of transition amplitudes, Eqs. (5.6) and (5.7) can be represented as:

$$a(u, d'; W^-) = a(u, d; W^-) \cos \theta_c + a(u, s; W^-) \sin \theta_c = a_w \qquad (5.8)$$

and

$$a(c, s'; W^-) = -a(c, d; W^-) \sin \theta_c + a(c, s; W^-) \cos \theta_c = a_w . \qquad (5.9)$$

In addition one has the relations:

$$a(u, s'; W^-) = -a(u, d; W^-) \sin \theta_c + a(u, s; W^-) \cos \theta_c = 0 \qquad (5.10)$$

and

$$a(c, d'; W^-) = a(c, d; W^-) \cos \theta_c + a(c, s; W^-) \sin \theta_c = 0. \qquad (5.11)$$

Equations (5.8) and (5.9) indicate that it is the d' and s' quarks which interact with the u and c quarks, respectively, with the full strength a_w. These equations for quarks correspond to Eq. (5.4) for leptons. Similarly, Eqs. (5.10) and (5.11) for quarks correspond to Eq. (5.5) for leptons. The former equations do not yield zero amplitudes because there exists some additive quantum number (analogous to muon lepton number) which is required to be conserved. This lack of a selection rule indicates that the notion of weak isospin symmetry for the doublets (u, d') and (c, s') is *dubious*: the assumption of the 'partially conserved' strangeness additive quantum number S leads to the incorrect quark-mixing theory of Cabibbo (see Section 6.2).

Equations (5.8) and (5.9) give:

$$a(u, d; W^-) = a_w \cos \theta_c , \quad a(u, s; W^-) = a_w \sin \theta_c . \qquad (5.12)$$

Thus in the two generation approximation of the SM, transitions involving $d \to u + W^-$ proceed with a strength proportional to $a_w^2 \cos^2 \theta_c \approx 0.95 a_w^2$, while transitions involving $s \to u + W^-$ proceed with a strength proportional to $a_w^2 \sin^2 \theta_c \approx 0.05 a_w^2$, as required by experiment.

In the SM, both the fundamental strong nuclear force and the fundamental electromagnetic force are described in terms of an $SU(3)$ and a $U(1)$ local gauge theory, respectively. On the other hand, the fundamental CC weak nuclear force, mediated by the massive W bosons has experienced difficulty in being described by an appropriate *local* gauge theory, since a local gauge theory normally requires the mediating particles to be *massless*.

Two approaches have been attempted to describe the CC weak nuclear force in an $SU(2)$ gauge theory within a weak isospin space: (1) in terms of weak isospin doublets of both leptons and quarks, and (2) in terms of a so-called 'electroweak theory', which relates the electromagnetic and the CC weak nuclear forces. Any approach to describing the CC weak nuclear force was required to account for their *universality* and also that this force violated parity.

As mentioned in Chapter 4, the first attempt at a theory of the weak nuclear force involved in the β-decay of a neutron into a proton was made by Fermi in 1934. Fermi considered the neutron β-decay as a process analogous to that of an electromagnetic transition, the electron-antineutrino pair playing the role of the emitted photon, with the process taking place at a single spacetime point, corresponding to the short-range nature of the underlying weak nuclear force.

Fermi described the β-decay process in terms of two interacting vector currents, analogous to the Dirac electromagnetic current:

$$j_{em}^{\mu} = \bar{\psi}\alpha_{\mu}\psi, \tag{5.13}$$

where ψ is the electron field and α_{μ} are Dirac matrices described in Section 3.3, so that the matrix element describing the process could be written as:

$$M = \frac{F}{\sqrt{2}}j_1^{\mu}j_2^{\mu}, \tag{5.14}$$

where F is the Fermi weak coupling constant and j_1^{μ} and j_2^{μ} are given by

$$j_1^{\mu} = \bar{\psi}_p\alpha_{\mu}\psi_n, \quad j_2^{\mu} = \bar{\psi}_{\nu}\alpha_{\mu}\psi_e. \tag{5.15}$$

Unfortunately, this four-fermion point contact model failed to describe later experimental data of CC weak nuclear interaction processes. This led to a generalization of the currents in the Fermi model. In addition to the vector currents involving α_{μ}, scalar, tensor, axial vector and pseudoscalar currents were introduced into the matrix elements describing the CC weak nuclear interaction processes. This generalization allowed all the available β-decay data at the time to be described. However, there still remained some outstanding problems and later more data, which could not be understood using the generalized Fermi model.

First, pion decay, such as

$$\pi^- \to \mu^- + \bar{\nu}_{\mu}, \tag{5.16}$$

could not be interpreted within the model until the quark structure of hadrons was proposed [3, 4, 6, 7], so that the above pion decay could be understood as the four-fermion transition:

$$\pi^- \equiv (d + \bar{u}) \to \mu^- + \bar{\nu}_{\mu}. \tag{5.17}$$

Second, it was noted that for sufficiently high energies (≈ 300 GeV) the matrix element M of Eq. (5.14) leads to a cross section for the scattering of two fermions, which violates the unitarity condition associated with the partial wave amplitudes.

It was natural to postulate the existence of a weak analogue of the photon, termed the intermediate vector boson, as was suggested by Klein in 1938. Thus a propagator, corresponding to a massive intermediate vector boson (called the W boson), mediating the CC weak nuclear interaction, was introduced into the matrix element M. For low energies, this gives:

$$M = \frac{a_w^4}{8M_W^2} j_1^\mu j_2^\mu , \qquad (5.18)$$

where M_W is the mass of the W boson and a_w is the universal CC weak nuclear force transition amplitude, so that equating Eqs. (5.14) and (5.18) M_W can be related to the Fermi weak coupling constant:

$$\frac{F}{\sqrt{2}} = \frac{a_w^4}{8M_W^2} . \qquad (5.19)$$

Thus, although the insertion of a massive boson propagator into the matrix element M did not fully resolve the unitarity problem [22], it does allow the boson mass to be estimated from Eq. (5.19) using values of F and a_w obtained from experiment: the estimated mass of the W boson is $M_W \approx 80$ GeV. This is very close to the measured mass of the W boson discovered in 1983.

Third, the 1957 discovery of parity violations in CC weak nuclear interaction processes was in contradiction with the original Fermi model, which only involved vector currents. This led to two new hypotheses: (1) the two-component neutrino in 1957; (2) the CC weak nuclear interaction involves only left-handed fermions and right-handed antifermions in 1958.

The two-component neutrino hypothesis requires the neutrino to be massless. In this case the neutrino will exist in a state of definite helicity. Helicity is the projection of spin along the direction of motion and spin-$\frac{1}{2}$ particles such as the neutrino occur with helicity $\pm\frac{1}{2}$, corresponding to spin projection parallel (called right-handed) or antiparallel (called left-handed) to the direction of motion. In 1958, the helicity of the neutrino participating in a CC weak nuclear interaction was measured by Maurice Goldhaber (1911-2011) and collaborators and was found to be negative and the neutrino left-handed. At the time this was taken as confirmation of the two-component hypothesis. However, in recent years, evidence has been found [23] that neutrinos have mass, albeit very small. Thus the left-handed nature of neutrinos was attributed to the CC weak nuclear interaction rather than to the neutrinos themselves, i.e. it arises as a consequence of the second hypothesis above.

If the second hypothesis is adopted, the matrix element describing the β-decay weak nuclear interaction processes may be written as in Eq. (5.14) but now the interacting currents become

$$j_1^\mu = \bar{\psi}_p \bar{\Gamma} \alpha_\mu \Gamma \psi_n = \bar{\psi}_p \alpha_\mu \Gamma \psi_n \,, \tag{5.20}$$

and

$$j_2^\mu = \bar{\psi}_e \alpha_\mu \Gamma \psi_\nu \,, \tag{5.21}$$

since

$$\bar{\Gamma} = \frac{1}{2}(1 + \alpha_5) \,, \quad \bar{\Gamma} \alpha_\mu = \alpha_\mu \Gamma \,, \quad \Gamma^2 = \Gamma = \frac{1}{2}(1 - \alpha_5) \,, \tag{5.22}$$

and

$$\alpha_5 = \begin{pmatrix} -1_2 & 0_2 \\ 0_2 & 1_2 \end{pmatrix} . \tag{5.23}$$

Here the presence of the projection operator Γ ensures that only the left-handed components of the fermion fields are involved and since $\bar{\Gamma}\Gamma = 0$ that any scalar, tensor and pseudoscalar interactions are forbidden.

Moreover, it should be noted that within the framework of the GM (see Chapter 6) the electron neutrino is actually *predicted* to be left-handed, although it is considered to be a composite particle having a small mass, so that the above remains a satisfactory theoretical description.

Indeed, as mentioned in Section 4.3, assuming a triplet of three vector bosons, W^+, W^0 and W^-, in a weak isospin space, Bludman showed that the V-A interaction of Sudarshan and Marshak, which explained in terms of only CC weak nuclear forces all the observed parity violations predicted by Lee and Yang to occur in weak nuclear interactions, could be described by a left-handed $SU(2)_L$ *global* gauge theory, rather than a left-handed $SU(2)_L$ *local* gauge theory.

Such a left-handed $SU(2)_L$ global gauge theory is consistent with all the various observations and assumptions of the weak isospin theory: the left-handed nature of the CC weak nuclear force, the observed parity violations and the massive nature of the vector W bosons mediating the CC weak nuclear force. The main implication of the above nature of the CC weak nuclear force is that this force is *not* a fundamental force, as is conventionally assumed in the SM.

Contrary to the above implication concerning the likely nature of the CC weak nuclear force, in 1961 Glashow proposed that the CC weak nuclear force, considered as a fundamental force, could be associated with the fundamental electromagnetic force, if the overall symmetry was extended to a $U(1) \times SU(2)_L$ local gauge theory. Glashow's model involved both a triplet weak isospin ($\mathbf{i} = 1$) of vector bosons (W^+, W^0, W^-) and a singlet weak isospin ($\mathbf{i} = 0$) vector boson B^0. The two neutral bosons 'mixed' in such a way that they produced a massive Z^0 boson and the massless photon (γ):

$$\gamma = B^0 \cos\theta_W + W^0 \sin\theta_W , \qquad (5.24)$$

$$Z^0 = -B^0 \sin\theta_W + W^0 \cos\theta_W , \qquad (5.25)$$

where θ_W is the electroweak mixing angle.

The weak nuclear interactions, mediated by the massive W^+, W^- and Z^0 vector bosons between all the elementary particles of the SM, fall into two classes: (i) the CC weak nuclear interactions involving the W^+ and W^- bosons and (ii) the neutral weak nuclear interactions involving the Z^0 boson. The CC weak nuclear force, by assumption was acting exclusively on left-handed particles and right-handed antiparticles, was described by a left-handed $SU(2)_L$ local gauge theory. On the other hand, the neutral weak nuclear force acted upon both left-handed and right-handed particles and antiparticles, similar to the electromagnetic force.

This 'electroweak theory' was a major step towards understanding the so-called 'electroweak connection': *the charge preserving weak nuclear force is completely fixed by the electromagnetic force and the CC weak nuclear force* [24, 25]. In this sense, it should be noted that the electromagnetic force involving the neutral photons and the weak nuclear forces involving the charged W bosons and the neutral Z bosons are *related* but are *not* strictly unified, since the relationship involves two independent coupling constants, electric charge and weak charge (see Chapter 11).

This proposal of Glashow was a second attempt, following the earlier one of Schwinger in 1957, to 'unify' the electromagnetic and weak nuclear forces. However, it also suffered from the fact that the large masses of the W^+, Z^0 and W^- bosons had to be inserted into the theory by hand. Moreover, the origin of the masses of the W and Z bosons remained unclear, especially since it seemed necessary for the gauge bosons to be massless, in order to maintain local gauge invariance.

This mass problem was resolved by Steven Weinberg (1933-) and Abdus Salam (1926-1996), who independently employed the idea of *spontaneous symmetry breaking* involving the so-called Higgs mechanism developed by Peter Higgs (1929-), Robert Brout (1928-2011) and François Englert (1932-). In this way the W and Z bosons acquire mass and the photon remains massless.

Spontaneous symmetry breaking occurs when the ground state, vacuum state or equilibrium state of a physical system does *not* share the underlying symmetries of the theory. Such symmetry breakings occur in condensed matter physics: e.g. when a ferromagnet is cooled through its Curie temperature it acquires a magnetization in one direction or its inverse direction. This is described as a *spontaneous* change, since one cannot predict in advance which direction will occur.

The above treatment of the electromagnetic and the weak nuclear forces in terms of a $U(1) \times SU(2)_L$ local gauge theory, involving the Higgs mechanism to provide the large masses of the W and Z bosons, has become known as the Glashow-Weinberg-Salam (GWS) model and forms one of the cornerstones of the SM. The model gives the relative masses of the W and Z bosons in terms of θ_W:

$$M_W = M_Z \cos \theta_W . \tag{5.26}$$

In 1971, Gerard 't Hooft (1946-) showed that the GWS model of the electroweak forces was *renormalizable* and this self-consistency of the theory led to its general acceptance. Finally in 1983 both the W and Z particles, with approximately the expected masses, were discovered [7] by Carlo Rubbia (1934-), Simon van der Meer (1925-2011) and collaborators, thereby confirming the GWS model.

However, not everyone is convinced that the GWS model is correct, since the SM upon which it is based is considered to be *incomplete* in the sense that it provides little physical understanding of several empirical observations such as (1) the existence of three generations of leptons and quarks, which apart from mass have similar properties; (2) the mass hierarchy of elementary particles, which form the basis of the SM; (3) the nature of the gravitational force; (4) the SM does not provide a *unified* origin of mass, contrary to Einstein's 1905 conclusion [11]; etc. Furthermore, there are three additional problems with the SM.

First, is the classification of the elementary particles of the SM, employing a diverse complicated scheme of additive quantum numbers, some of which are not conserved in CC weak nuclear force processes; and at the same time failing to provide any physical basis for this scheme. Second, concerns the dubious method that the SM employs to accommodate the universality of the CC weak nuclear forces. Third, concerns the origin of mass: in the SM the masses of hadrons arise mainly from the energy content of their constituent quarks and gluons, in agreement with Einstein's 1905 conclusion; on the other hand the masses of the elementary particles, the leptons, the quarks and the W and Z bosons are interpreted quite differently, arising from the existence of the so-called Higgs field [4]. These problems will be discussed further in the following chapter.

Chapter 6

Generation Model of Particle Physics

6.1 Introduction

In this chapter, the current formulation of the Generation Model of particle physics (GM) [5, 26, 27] will be presented as an alternative to the SM. The GM has been developed, primarily to overcome the deficiencies of the SM, during the last two decades by Brian Robson (1934-), the author of this book.

In the previous chapter, it was emphasized that the current SM, while enjoying considerable success in describing the interactions of leptons and the multitude of hadrons with each other, as well as the decay modes of the unstable leptons and hadrons, is generally considered to be *incomplete* in the sense that it provides little understanding of several empirical observations.

This inability of the SM to provide an understanding of such important empirical observations has become very frustrating to particle physicists, especially the experimentalists, who find that all their other data are adequately described by the SM. In order to comprehend this dilemma and to progress beyond the SM it is necessary to reconsider the basic assumptions upon which the SM is built.

An excellent analogy of the SM situation is the Ptolemaic model of the Universe, based upon a stationary Earth at the center surrounded by a rotating system of crystal spheres refined by the addition of epicycles (small circular orbits) to describe the peculiar movements of the planets around the Earth. While the Ptolemaic model yielded an excellent description, it is a complicated diverse scheme for predicting the movements of the Sun, Moon, planets and the stars around a stationary Earth and unfortunately provides no understanding of these complicated movements.

Progress in understanding the Universe was only made when the Ptolemaic model was replaced by the Copernican-Keplerian model, in which the Earth moved like the other planets around the Sun, and Newton discovered his universal law of gravitation to describe the approximately elliptically planetary orbits. Indeed, it was only by removing the incorrect assumptions of a stationary Earth and that celestial bodies moved in 'divine circles' that progress beyond the Ptolemaic model was achieved.

In the previous chapter, three significant problems with the SM were highlighted. First, the classification of the elementary leptons and quarks in terms of additive quantum numbers is both nonunified and complex, and in addition some additive quantum numbers are not conserved in the CC weak nuclear interactions. Second, the dubious method involving the Cabibbo technique to describe the universality of the CC weak nuclear interactions. This leads to a sharing of the CC weak nuclear interactions between quarks of different generations without the existence of a conserved additive quantum number, corresponding to the lepton numbers within the lepton sector. Third, the assumption that the CC weak nuclear force is a fundamental force described by a local gauge theory, contrary to the usual requirement that the mediating massive W bosons should be massless.

The GM replaces the above three dubious assumptions of the SM by three different and simpler assumptions. These are: (1) the assumption of a _simpler unified_ classification scheme of leptons and quarks; (2) the assumption that the _mass eigenstate quarks_ form weak isospin doublets and that hadrons are composed of _weak eigenstate quarks_ and (3) the assumption that the CC weak nuclear force is _not_ a fundamental force.

6.2 Historical Development: 2002 GM

The development of a successful alternative model of particle physics to the SM commenced in 2002 [28]. This paper entitled "A Generation Model of the Fundamental Particles", was the first publication to mention the alternative model as the Generation Model (GM).

The paper describes a new simpler and unified classification of the elementary leptons and quarks of the SM in terms of only _three_ additive quantum numbers, charge (Q), particle number (p) and generation quantum number (g), rather than the more complicated and nonunified scheme of the SM involving the _nine_ additive quantum numbers, charge (Q), lepton number (L), muon lepton number (L_μ), tau lepton number (L_τ), baryon number (A), strangeness (S), charm (C), bottomness (B) and topness (T), (see Table 5.1).

Table 6.1 displays a possible set of the three additive quantum numbers, Q, p and g that were found to be conserved in all interactions involving leptons and quarks. The charge quantum number (Q) was introduced into the SM to describe the conservation of electric charge: in this 2002 GM, the charge quantum number serves the same purpose. The particle quantum number (p) replaces both the baryon quantum number (A) of quarks and the lepton quantum number (L) of leptons in the SM, so that $p = \frac{1}{3}$ for quarks and $p = -1$ for leptons, essentially in agreement with the corresponding quantum numbers of the SM. The generation quantum number (g) replaces the remaining six additive quantum numbers of the SM: L_μ, L_τ, S, C, B and T. In this way, it distinguishes the different generations and lends its name to the model itself. The choice of $g = 0$ for the first generation of two leptons and two quarks, (ν_e, e^-, u, d), is natural, since these are the constituents of ordinary matter for which the remaining six additive quantum numbers are all zero. The choice of $g = -1$ for the second generation, (ν_μ, μ^-, c, s), conforms to the choice of $S = -1$ for the strange quark in the SM, while the choice of $g = 1$ for the third generation, (ν_τ, τ^-, t, b), simply reflects the triplet nature of the three generations: nature seems to require only three generations.

Particle	Q	p	g	Particle	Q	p	g
ν_e	0	-1	0	u	$\frac{2}{3}$	$\frac{1}{3}$	0
e^-	-1	-1	0	d	$-\frac{1}{3}$	$\frac{1}{3}$	0
ν_μ	0	-1	-1	c	$\frac{2}{3}$	$\frac{1}{3}$	-1
μ^-	-1	-1	-1	s	$-\frac{1}{3}$	$\frac{1}{3}$	-1
ν_τ	0	-1	1	t	$\frac{2}{3}$	$\frac{1}{3}$	1
τ^-	-1	-1	1	b	$-\frac{1}{3}$	$\frac{1}{3}$	1

Table 6.1. 2002 GM additive quantum numbers for leptons and quarks.

In the 2002 GM the force particles are allocated the additive quantum numbers shown in Table 6.2. In particular, each force particle has $p = g = 0$.

Particle	Q	p	g
γ	0	0	0
W^-	−1	0	0
Z^0	0	0	0
W^+	1	0	0
gluon	0	0	0

Table 6.2. 2002 GM additive quantum numbers for force particles.

To summarize, this initial 2002 form of the GM provided a simple unified classification scheme of the leptons and quarks in terms of only three additive quantum numbers, Q, p and g. Furthermore, each of these additive quantum numbers was also required to be *conserved* in all interactions, unlike the S, C, B and T additive quantum numbers of the SM that are not strictly conserved in the weak nuclear interactions.

Thus the initial 2002 GM overcame the first major problem of the SM, involving a complicated nonunified classification scheme that presented a major stumbling block to the development of a composite model of the twelve elementary leptons and quarks of the SM. Indeed many such composite models had been proposed prior to 1983 but these had met with little success, primarily in my view because it was difficult to relate the composite models to the complicated nonunified classification scheme of the relatively successful SM.

Furthermore, the 2002 GM also replaced the second dubious assumption of the SM that the u and c quarks formed weak isospin doublets with the weak nuclear force eigenstate quarks, d' and s', respectively (see Eqs. (5.6) and (5.7)), by placing the quark mixing in the quark states (wave functions) rather than in the CC weak nuclear interactions as proposed by Cabibbo.

In the 2002 GM, it is postulated that the mass eigenstate quarks of the same generation, e.g. (u, d), form weak isospin doublets and couple with the full strength of the CC weak nuclear interaction, a_w, like the lepton doublets, e.g. (ν_e, e^-). Contrary to the SM, the GM requires that there is no coupling between mass eigenstate quarks from different generations. This latter requirement corresponds to the conservation of the generation quantum number g in the CC weak nuclear force processes.

Second, the 2002 GM postulated that hadrons are composed of weak eigenstate quarks such as d' and s', rather than the corresponding mass eigenstate quarks, d and s, as in the SM. Thus, in terms of transition amplitudes, the 2002 GM postulated that:

$$a(u, d; W^-) = a(c, s; W^-) = a_w, \qquad (6.1)$$

and

$$a(u, s; W^-) = a(c, d; W^-) = 0. \qquad (6.2)$$

Equations (6.1) and (6.2) are the analogues of Eqs. (5.4) and (5.5) for leptons. These transition amplitudes establish a close lepton-quark parallelism with respect to weak isospin symmetry.

The reduced transition probabilities for neutron and Λ^0 β-decays relative to muon decay, assuming that the neutron and Λ^0 baryon are composed of weak eigenstate quarks, u, d' and s', are given by the following sequential transitions. For neutron β-decay one has:

$$d' \to u + W^-, \quad W^- \to e^- + \bar{\nu}_e, \qquad (6.3)$$

and assuming the two generation approximation, i.e. only Cabibbo quark mixing, the primary transition has the amplitude $a(u, d'; W^-)$ given by

$$a(u, d'; W^-) = a(u, d; W^-)\cos\theta_c + a(u, s; W^-)\sin\theta_c = a_w\cos\theta_c, \quad (6.4)$$

using Eqs. (6.1) and (6.2). This gives the same transition probability for neutron β-decay ($a_w^4 \cos^2\theta_c$) relative to muon decay (a_w^4) as the SM. Similarly, Λ^0 β-decay is to be treated as

$$s' \to u + W^-, \quad W^- \to e^- + \bar{\nu}_e. \qquad (6.5)$$

In this case the primary transition has the amplitude $a(u, s'; W^-)$ given by

$$a(u, s'; W^-) = -a(u, d; W^-)\sin\theta_c + a(u, s; W^-)\cos\theta_c = -a_w\sin\theta_c. \qquad (6.6)$$

Thus Λ^0 β-decay has the same transition probability ($a_w^4 \sin^2\theta_c$) relative to muon decay (a_w^4) as that given by the SM.

One of the essential differences between the 2002 GM and the SM lies in the treatment of quark-mixing, which is required to account for the universality of the CC weak nuclear interactions.

In the SM, it is assumed that the pairs of fermions (ν_e, e^-), (ν_μ, μ^-), (ν_τ, τ^-), (u, d'), (c, s') and (t, b'), where d', s' and b' are CKM mixed states of the physical quarks, d, s and b, form weak isospin doublets. This implies that the CC weak nuclear interactions, involving the members of each pair and a charged W boson, are identical, i.e. the universal CC weak nuclear interaction. This means that the universal CC weak nuclear interaction, in the case of the pair (u, d'), is *shared* between the components of d'.

In the 2002 GM, it is assumed that the pairs of fermions (ν_e, e^-), (ν_μ, μ^-), (ν_τ, τ^-), (u, d), (c, s) and (t, b), form weak isospin doublets so that the members of each pair and a charged W boson are the universal CC weak nuclear interaction. However, in a given CC weak nuclear interaction, the 2002 GM assumes that hadrons are composed of CKM mixed quarks, so that the transition probability for any CC weak nuclear process in the 2002 GM is equivalent to the corresponding one in the SM.

Hadrons composed of mixed-quark states might seem to suggest that the electromagnetic and strong nuclear interaction processes between mass eigenstate hadron components are not consistent with the fact that weak nuclear interaction processes occur between weak eigenstate quarks. However, since the electromagnetic and the strong nuclear interactions are *flavor independent*: the down, strange and bottom quarks carry the same electric and color charges so that the weak eigenstate quarks have the same magnitude of electric and color charge as the mass eigenstate quarks. Therefore, it becomes apparent that the weak nuclear interaction is the *only* interaction in which the quark-mixing phenomenon can be detected.

The different treatment of quark mixing in the 2002 GM has a major advantage: it allowed a conserved additive quantum number, the generation quantum number, to be allotted to the physical quarks, unlike the S, C, B and T quantum numbers of the SM. Furthermore, the existence of mixed-quark states in hadrons implied that strange quarks are present in hadrons such as neutrons and protons. Such other differences between the 2002 GM and the SM will be discussed later.

6.3   Historical Development: 2011 GM

The second major step in the development of an alternative model to the SM took place during the period 2003-2011 and resulted in the 2011 GM [5], which is a composite model of the leptons and quarks of the SM. Indeed, during the late 20th century, numerous such models had been proposed [29]. The underlying reason for this was that the twelve elementary particles of the SM, the six leptons and the six quarks, was considered to be too many basic particles. In addition there was considerable indirect evidence that the leptons and quarks were probably composite particles.

First, the electric charges of the electron and proton are opposite in sign but are *exactly* equal in magnitude so that atoms with the same number of electrons and protons are neutral. Consequently, in a proton consisting of quarks, the electric charges of the quarks are intimately related to that of the electron: in fact, the up quark has electric charge $Q = +\frac{2}{3}$ and the down quark has electric charge $Q = -\frac{1}{3}$, if the electron has electric charge $Q = -1$. These relations are readily comprehensible if leptons and quarks are composed of the same kinds of particles.

Second, in the SM the six leptons and the six quarks can be grouped into three *generations*: (i) (e^-, ν_e, u, d), (ii) (μ^-, ν_μ, c, s) and (iii) (τ^-, ν_τ, t, b). Each generation contains particles which have similar properties. Corresponding to the electron, e^-, the second and third generations include the muon, μ^-, and the tau particle, τ^-, respectively. Each generation contains a neutrino associated with the corresponding leptons: the electron neutrino, ν_e, the muon neutrino, ν_μ, and the tau neutrino, ν_τ. In addition, each generation contains a quark with $Q = +\frac{2}{3}$, the up quark, u, the charmed quark, c and the top quark, t, and a quark with electric charge $Q = -\frac{1}{3}$, the down quark, d, the strange quark, s and the bottom quark, b. Each pair of leptons and quarks of the same generation are connected by isospin symmetries, otherwise the grouping into the three families is according to increasing mass of the corresponding family members. The existence of three repeating patterns suggests strongly that the members of each generation are composite particles, analogous to the composite elements in the same vertical column of the Mendeleev periodic table that have similar chemical properties.

As mentioned in Chapter 1, in 2001 Veltman stated during a public lecture that the *greatest puzzle of elementary particle physics was the occurrence of three families of elementary particles that have the same properties except for mass* in the SM. During 2001 I concluded that a basic problem with the SM was its *nonunified* classification of its elementary leptons and quarks: this presented a major stumbling block for the development of a composite model of these particles. In 2002 I had developed the 2002 GM, which did not suffer from the same stumbling block, since this model possessed a *unified* classification scheme of the leptons and quarks. Consequently, in 2003 I set out to develop a composite version of the 2002 GM. After considerable contemplation, this led to the 2011 GM, which was capable of describing the three families of the SM, in terms of a composite model of the leptons and quarks.

The 2011 GM description of the first generation of leptons and quarks in the SM is based partly upon the two-particle models of Haim Harari (1940-) and Michael Shupe. Both models proposed in 1979 are very similar, and arguably provide an economical and impressive description of the first generation of leptons and quarks, and their antiparticles in the SM. Both models treat the leptons and quarks as composites of only two kinds of massless spin-$\frac{1}{2}$ particles, which Harari named 'rishons' from the Hebrew word for primary. This name was adopted for the constituents of the leptons and quarks in the development of the 2011 GM.

In the Harari-Shupe Model (HSM), two kinds of massless spin-$\frac{1}{2}$ elementary particles, T and V rishons, and their corresponding antiparticles are employed to construct the leptons and quarks: (i) a T-rishon with electric charge $Q = +\frac{1}{3}$ and (ii) a V-rishon with $Q = 0$. Their antiparticles are a $\bar{T}$-antirishon with $Q = -\frac{1}{3}$ and a $\bar{V}$-antirishon with $Q = 0$, respectively. In addition each lepton and quark is assumed to be composed of three rishons or three antirishons.

Particle	*structure*	Q
e^+	TTT	$+1$
u	TTV, TVT, VTT	$+\frac{2}{3}$
$\bar{d}$	TVV, VTV, VVT	$+\frac{1}{3}$
ν_e	VVV	0
$\bar{\nu}_e$	$\bar{V}\bar{V}\bar{V}$	0
d	$\bar{T}\bar{V}\bar{V}, \bar{V}\bar{T}\bar{V}, \bar{V}\bar{V}\bar{T}$	$-\frac{1}{3}$
$\bar{u}$	$\bar{T}\bar{T}\bar{V}, \bar{T}\bar{V}\bar{T}, \bar{V}\bar{T}\bar{T}$	$-\frac{2}{3}$
e^-	$\bar{T}\bar{T}\bar{T}$	-1

Table 6.3. HSM of first generation of leptons and quarks.

Table 6.3 shows the proposed structures of the first generation of leptons and quarks in the HSM. Essentially, the HSM describes the electric charge character of the first generation of particles. It should be noted that no composite particle involves a combination of rishons and antirishons, as emphasized by Shupe. Both Harari and Shupe noted that quarks contained mixtures of the two types of rishons, whereas leptons did not. They also experienced some difficulty in understanding the color charges of quarks. Initially it was considered that the three different color charges of quarks may be related to different internal arrangements of the rishons within a quark. For example, the structure of the u quark (see Table 6.3) indicates three possible different arrangements: TTV, TVT and VTT so that perhaps these corresponded to the three different color charges, red, green and blue of the u quark. However, no structural mechanism was suggested to account for this.

During 1980-82 several composite models, based upon the HSM, were proposed to describe the color charge of quarks in terms of the rishon model. These models involved *two* color-type $SU(3)$ local gauge symmetries, namely $SU(3)_H \times SU(3)_C$ at the rishon level. First, one had an additional strong hypergluon interaction, corresponding to the $SU(3)_H$ symmetry, mediated by massless *hypergluons*, which was responsible for binding rishons, carrying a hypercolor charge, together to form hypercolorless leptons and quarks. This implied that each lepton and quark was required to be composed of three rishons. Second, one retained the strong chromodynamic interaction of the SM, corresponding to the $SU(3)_C$ symmetry, mediated by massless gluons, which is responsible for binding color charged quarks together to form colorless baryons or mesons.

In each of the proposed composite models, both the new hypercolor interaction and the original color interaction are assumed to be similar but exist independently of one another, so that the original rishon model loses some of its economical description. This two-fold color charge complication, together with the considerable difficulty involved in relating the HSM to the nonunified classification scheme of the SM, probably caused the proposed composite models to flounder without any significant success. Indeed, none of these models provided a satisfactory understanding of the second and third generations of the SM.

In order to overcome some of the deficiencies of the original HSM, the two-rishon model was extended, within the framework of the 2011 GM, in several ways.

First, it was noted that Eq. (4.12) could be written in terms of the additive quantum numbers, Q, p and g, in the form:

$$Q = I_3 + \frac{1}{2}(p + g), \tag{6.7}$$

since in the 2002 GM, the particle number p replaces A and the generation quantum number replaces S, C, B and T. This suggested the existence of an $SU(3)_f$ flavor symmetry underlying the substructure of leptons and quarks. Consequently, a third kind of rishon, U, was introduced into the 2011 GM, which treats leptons and quarks as composites of *three* kinds of massless spin-$\frac{1}{2}$ rishons and/or their antiparticles, although the U-rishon is only involved in the second and third generations.

Second, in the 2011 GM, each rishon is allotted both a particle number p and a generation quantum number g. Table 6.4 shows the three additive quantum numbers allotted to the three kinds of rishons. It should be noted that for each rishon additive quantum number N, the corresponding antirishon has the additive quantum number $-N$.

rishon	Q	p	g
T	$+\frac{1}{3}$	$+\frac{1}{3}$	0
V	0	$+\frac{1}{3}$	0
U	0	$+\frac{1}{3}$	-1

Table 6.4. 2011 GM additive quantum numbers for rishons.

Historically, the term 'particle' defines matter that is naturally occurring, i.e. electrons, neutrons and protons. In the 2011 GM it was convenient to define a matter 'particle' to have $p > 0$, with the antiparticle having $p < 0$. This definition of a matter particle leads to a modification of the HSM structures of the leptons and quarks which comprise the first generation. Essentially, the roles of the V-rishon and its antiparticle $\bar{V}$ are interchanged in the 2011 GM compared with the HSM. Table 6.5 gives the 2011 GM structures for the first generation of leptons and quarks. The particle number p is clearly given by $\frac{1}{3}$(number of rishons − number of antirishons). Thus the u-quark has $p = +\frac{1}{3}$, since it contains two T-rishons and one $\bar{V}$-antirishon. It should be noted that it is essential for the u-quark to contain a $\bar{V}$-antirishon ($p = -\frac{1}{3}$) rather than a V-rishon ($p = +\frac{1}{3}$) to obtain a value of $p = +\frac{1}{3}$, corresponding to baryon number $A = +\frac{1}{3}$ in the SM.

particle	*structure*	Q	p	g
e^+	TTT	$+1$	$+1$	0
u	$TT\bar{V}$	$+\frac{2}{3}$	$+\frac{1}{3}$	0
$\bar{d}$	$T\bar{V}\bar{V}$	$+\frac{1}{3}$	$-\frac{1}{3}$	0
ν_e	$\bar{V}\bar{V}\bar{V}$	0	-1	0
$\bar{\nu}_e$	VVV	0	$+1$	0
d	$\bar{T}VV$	$-\frac{1}{3}$	$+\frac{1}{3}$	0
$\bar{u}$	$\bar{T}\bar{T}V$	$-\frac{2}{3}$	$-\frac{1}{3}$	0
e^-	$\bar{T}\bar{T}\bar{T}$	-1	-1	0

Table 6.5. 2011 GM of first generation of leptons and quarks.

In the 2011 GM, no significance is attached to the ordering of the T-rishons and the $\bar{V}$-antirishons, so that, e.g. the structures $TT\bar{V}$, $T\bar{V}T$ and $\bar{V}TT$ for the u-quark are considered to be equivalent. The concept of color is treated differently in the 2011 GM: it is assumed that all three rishons, T,

V and U carry a color charge, red, green or blue, while their antiparticles carry an anticolor charge, antired, antigreen or antiblue. The 2011 GM assumes a strong color-type interaction corresponding to a local gauged $SU(3)_C$ symmetry (analogous to QCD) mediated by massless neutral spin-1 *hypergluons*, to be responsible for binding rishons and antirishons together to form colorless leptons and colored quarks.

In the 2011 GM each lepton of the first generation (see Table 6.5) is assumed to be colorless, consisting of three rishons (or antirishons), each with a different color charge (or anticolor charge), analogous to the baryons (or antibaryons) of the SM. These leptons are built out of T-rishons and V-rishons or their antiparticles $\bar{T}$ and $\bar{V}$, all of which have generation quantum number $g = 0$.

It is envisaged that each lepton of the first generation exists in an antisymmetrical three-particle colorless state, which physically assumes a quantum mechanical triangular distribution of the three differently colored identical rishons (or antirishons), since each of the three color interactions between pairs of rishons (or antirishons) is expected to be strongly attractive [30].

In the 2011 GM, it is assumed that each quark of the first generation is a composite of a colored rishon and a colorless rishon-antirishon pair, $(T\bar{V})$ or $(V\bar{T})$, so that the quarks carry a color charge. Similarly, the antiquarks are a composite of an anticolored antirishon and a colorless rishon-antirishon pair, so that the antiquarks carry an anticolor charge. The proposed structures of the quarks of the first generation require the composite quarks to have a color charge so that the dominant residual interaction between such quarks is essentially the same as that between rishons, and consequently these composite quarks behave very like the elementary quarks of the SM. In the 2011 GM the term 'hypergluon' is retained as the mediator of the strong color interaction, rather than the term 'gluon' employed in the SM, because it is the rishons rather than the quarks that carry an elementary color charge.

It is envisaged that each quark of the first generation will exist in essentially a linear distribution, since the interactions between the rishon-antirishon colorless pair and between the pair of rishons of different color charges are strongly attractive, while the remaining interaction between the rishon with a different color and the antirishon of the colorless rishon-antirishon pair is strongly repulsive.

In order to preserve the universality of the CC weak nuclear interaction processes involving first generation quarks, e.g. the transition

$d \to u + W^-$, it is assumed that the first generation quarks have the general color structures:

up quark : $T_C(T_{C'}\bar{V}_{\bar{C}'})$, down quark : $V_C(V_{C'}\bar{T}_{\bar{C}'})$, with $C' \neq C$. (6.8)

Thus a red u-quark and a red d-quark have the general color structures:

$$u_r = T_r(T_g\bar{V}_{\bar{g}} + T_b\bar{V}_{\bar{b}})/\sqrt{2}, \qquad (6.9)$$

and

$$d_r = V_r(V_g\bar{T}_{\bar{g}} + V_b\bar{T}_{\bar{b}})/\sqrt{2}, \qquad (6.10)$$

respectively. For $d_r \to u_r + W^-$, conserving color, one has the two transitions:

$$V_r V_g \bar{T}_{\bar{g}} \to T_r T_b \bar{V}_{\bar{b}} + V_r V_g V_b \bar{T}_{\bar{r}} \bar{T}_{\bar{g}} \bar{T}_{\bar{b}} \qquad (6.11)$$

and

$$V_r V_b \bar{T}_{\bar{b}} \to T_r T_g \bar{V}_{\bar{g}} + V_r V_g V_b \bar{T}_{\bar{r}} \bar{T}_{\bar{g}} \bar{T}_{\bar{b}}, \qquad (6.12)$$

which take place with equal probabilities. In these transitions, the W^- boson is assumed to be a three $\bar{T}$-antirishon and a three V-rishon colorless composite particle with additive quantum numbers $Q = -1$, $p = g = 0$. The corresponding W^+ boson has the structure $[T_r T_g T_b \bar{V}_{\bar{r}} \bar{V}_{\bar{g}} \bar{V}_{\bar{b}}]$, consisting of a colorless set of three T-rishons and a colorless set of three $\bar{V}$-antirishons with additive quantum numbers $Q = +1$, $p = g = 0$.

Particle	*structure*	Q	p	g
μ^+	$TTT\Pi$	$+1$	$+1$	±1
c	$TT\bar{V}\Pi$	$+\frac{2}{3}$	$+\frac{1}{3}$	±1
$\bar{s}$	$T\bar{V}\bar{V}\Pi$	$+\frac{1}{3}$	$-\frac{1}{3}$	±1
ν_μ	$\bar{V}\bar{V}\bar{V}\Pi$	0	-1	±1
$\bar{\nu}_\mu$	$VVV\Pi$	0	$+1$	±1
s	$\bar{T}VV\Pi$	$-\frac{1}{3}$	$+\frac{1}{3}$	±1
$\bar{c}$	$\bar{T}\bar{T}V\Pi$	$-\frac{2}{3}$	$-\frac{1}{3}$	±1
μ^-	$\bar{T}\bar{T}\bar{T}\Pi$	-1	-1	±1

Table 6.6. 2011 GM of second generation of leptons and quarks.

The rishon structures of the second generation particles are the same as the corresponding particles of the first generation plus the addition of a colorless rishon-antirishon pair, Π, where

$$\Pi = [(\bar{U}V) + (\bar{V}U)]/\sqrt{2}, \qquad (6.13)$$

which is a quantum mechanical mixture of $(\bar{U}V)$ and $(\bar{V}U)$, which have $Q = p = 0$ but $g = \pm 1$, respectively. In this way, the pattern for the first generation is repeated for the second generation. Table 6.6 gives the 2011 GM structures for the second generation of leptons and quarks.

It should be noted that for any given transition the generation quantum number is required to be conserved, although each particle of the second generation has two possible values of g. For example, the decay

$$\mu^- \to \nu_\mu + W^- , \tag{6.14}$$

at the rishon level may be written

$$\bar{T}\bar{T}\bar{T}\Pi \to \bar{V}\bar{V}\bar{V}\Pi + \bar{T}\bar{T}\bar{T}VVV , \tag{6.15}$$

which proceeds via the two transitions:

$$\bar{T}\bar{T}\bar{T}(\bar{U}V) \to \bar{V}\bar{V}\bar{V}(\bar{U}V) + \bar{T}\bar{T}\bar{T}VVV \tag{6.16}$$

and

$$\bar{T}\bar{T}\bar{T}(\bar{V}U) \to \bar{V}\bar{V}\bar{V}(\bar{V}U) + \bar{T}\bar{T}\bar{T}VVV , \tag{6.17}$$

which take place with equal probabilities. In each case, the additional colorless rishon-antirishon pair, $(\bar{U}V)$ or $(\bar{V}U)$, essentially acts as a spectator during the CC weak nuclear interaction process.

Particle	*structure*	Q	p	g
τ^+	$TTT\Pi\Pi$	$+1$	$+1$	$0, \pm 2$
t	$TT\bar{V}\Pi\Pi$	$+\frac{2}{3}$	$+\frac{1}{3}$	$0, \pm 2$
$\bar{b}$	$T\bar{V}\bar{V}\Pi\Pi$	$+\frac{1}{3}$	$-\frac{1}{3}$	$0, \pm 2$
ν_τ	$\bar{V}\bar{V}\bar{V}\Pi\Pi$	0	-1	$0, \pm 2$
$\bar{\nu}_\tau$	$VVV\Pi\Pi$	0	$+1$	$0, \pm 2$
b	$\bar{T}VV\Pi\Pi$	$-\frac{1}{3}$	$+\frac{1}{3}$	$0, \pm 2$
$\bar{t}$	$\bar{T}\bar{T}V\Pi\Pi$	$-\frac{2}{3}$	$-\frac{1}{3}$	$0, \pm 2$
τ^-	$\bar{T}\bar{T}\bar{T}\Pi\Pi$	-1	-1	$0, \pm 2$

Table 6.7. 2011 GM of third generation of leptons and quarks.

The rishon structures of the third generation particles are the same as the corresponding particles of the first generation plus the addition of two rishon-antirishon pairs, which are a quantum mechanical mixture of $(\bar{U}V)$ and $(\bar{V}U)$ and, as for the second generation, are assumed to be colorless and have $Q = p = 0$ but $g = \pm 1$. In this way the pattern of the first and second generation is continued for the third generation. Table 6.7 gives the 2011 GM structures for the third generation of leptons and quarks.

The rishon structure of the τ^+ particle is

$$TTT\Pi\Pi = TTT[(\bar{U}V)(\bar{U}V) + (\bar{U}V)(\bar{V}U) \qquad (6.18)$$
$$+ (\bar{V}U)(\bar{U}V) + (\bar{V}U)(\bar{V}U)]/2$$

and each particle of the third generation is a similar quantum mechanical mixture of $g = 0, \pm2$ components. The color structures of both second and third generation leptons and quarks have been chosen so that the CC weak nuclear interaction is universal. In each case, the additional colorless rishon-antirishon pairs, $(\bar{U}V)$ and/or $(\bar{V}U)$, essentially act as spectators during any CC weak nuclear interaction process. Again it should be noted that for any given transition the generation quantum number is required to be conserved, although each particle of the third generation now has three possible values of g. Furthermore, in the 2011 GM the three independent additive quantum numbers, charge Q, particle number p and generation quantum number g, which are conserved in all interactions, correspond to the conservation of each of the three kinds of rishons:

$$n(T) - n(\bar{T}) = 3Q, \qquad (6.19)$$

$$n(\bar{U}) - n(U) = g, \qquad (6.20)$$

$$n(T) + n(V) + n(U) - n(\bar{T}) - n(\bar{V}) - n(\bar{U}) = 3p, \qquad (6.21)$$

where $n(R)$ and $n(\bar{R})$ are the numbers of rishons and antirishons, respectively. Thus, the conservation of g in weak interactions is a consequence of the conservation of the three kinds of rishons $(T, V$ and $U)$, which also prohibits transitions between the third generation and the first generation via weak interactions even for $g = 0$ components of third generation particles.

6.4　Historical Development: 2019 GM

The third and most recent step in the development of an alternative model to the SM took place during 2018 and resulted in the 2019 GM [27], which is a simpler model than the 2011 GM, employing only *two*, rather than *three* kinds of rishons and their antiparticles.

The essential reason for introducing the U-rishon and its antiparticle $\bar{U}$-antirishon into the 2011 GM was to avoid annihilation processes, if the second and third generations of leptons and quarks involved colorless rishon-antirishon pairs $V\bar{V}$.

During 2018 I considered the physical nature of the U-rishon. I noted that while the T-rishon ($Q = +\frac{1}{3}$) and the V-rishon ($Q = 0$) are distinguishable by their electric charges, the U-rishon ($Q = 0$), which carries generation quantum number $g = -1$, is *not* distinguishable from the V-rishon by any physical characteristic: the generation quantum number is not associated with any physical property of the U-rishon.

Following further contemplation, it occurred to me plausible that the 2011 GM, employing three kinds of rishons and their antiparticles (all tacitly assumed to be in a 1s ground state) may be mimicking a *simpler* model, employing only two kinds of rishons and their antiparticles. Perhaps the U-rishon was simply a V-rishon existing in an excited state, i.e. a V^*-rishon, and its antiparticle was likewise a $\bar{V}^*$-antirishon.

The U-rishon and its antiparticle $\bar{U}$-antirishon only occur in the second and third generations of the 2011 GM, so that it was quite plausible that the V^*-rishon and the $\bar{V}^*$-antirishon could occupy a 2s state by analogy with atomic physics [31]. Furthermore, annihilation processes are avoided, since the excited-state V^*-rishon and also its excited-state antiparticle $\bar{V}^*$-antirishon will exist in an orthogonal quantum state to that of the V-rishon and the $\bar{V}$-antirishon in a 1s state.

To summarize: the 2011 GM, which describes the elementary particles of the SM as composite particles in terms of three kinds of rishons and their antiparticles, may be replaced by a simpler equivalent model employing only two kinds of rishons and their antiparticles. If the excited V^*-rishon occupies a 2s state in the composite systems describing the leptons and quarks of both the second and third generations, all the interactions involving the leptons and quarks within the framework of the simpler GM are *identical* to those of the 2011 GM.

This simpler 2019 GM is the current GM and henceforth in the following chapters will be referred to as simply *the* GM.

There is one additional important point to make concerning the composite versions of the GM: the building blocks of the GM are assumed to be *massless* spin-$\frac{1}{2}$ rishons and antirishons, which have intrinsic parity $+1$ and -1, respectively. This implies that all the composite leptons and quarks also have an intrinsic parity ± 1, depending upon the number of rishons and the number of antirishons comprising each composite particle, provided that it is assumed that each rishon and antirishon exists in an s state. Thus, e.g. the electron and the electron neutrino both have negative intrinsic parity and are left-handed particles, while the muon and the muon neutrino both have positive intrinsic parity and are right-handed antiparticles. Consequently, the right-handed electron and the right-handed electron neutrino, and similarly the left-handed muon and the left-handed muon neutrino, do not exist in the GM.

6.5 Fundamental and Residual Forces in GM

The GM recognizes only two fundamental forces in nature: (1) the usual electromagnetic force, mediated by massless neutral spin-1 photons between electrically charged particles and described by a $U(1)$ local gauge theory called QED (see Section 3.3) and (2) the strong chromodynamic force, mediated by massless neutral spin-1 hypergluons, acting between color charged rishons and/or antirishons and described by an $SU(3)$ local gauge theory called QCD (see Section 6.3).

The above two fundamental forces: the electromagnetic force and the chromodynamic force are essentially equivalent in both the GM and the SM, since the electric charge states and the color charge states of the composite quarks of the GM and the elementary quarks of the SM are essentially identical. The same is also the case of the residual strong nuclear force acting between the quark constituents of colorless hadrons, mediated by colorless mesons , which are sufficiently strong so that neutrons and protons are held together within atomic nuclei.

However, the substructure of the leptons and quarks in the GM implies *two* additional residual forces of the strong color forces binding together the rishon and/or antirishon constituents of the composite leptons and quarks. These are: (1) the CC weak nuclear force, mediated by massive spin-1 composite colorless W bosons (see Section 6.3), acting between the rishon and/or antirishon constituents of the fermions of the weak isospin doublets, such as (ν_e, e^-) and (u, d), and (2) the gravitational force, mediated by massless neutral spin-1 hypergluons, acting between the rishon and/or antirishon constituents of colorless particles such as electrons, protons and neutrons. Gravity will be discussed in more detail in Chapter 8.

As discussed in Section 6.3, the CC weak nuclear force is mediated by massive colorless W^+ and W^- bosons, composed of three $\bar{V}$-antirishons of different anticolor charges and three T-rishons of different color charges, and three V-rishons of different color charges and three $\bar{T}$-antirishons of different anticolor charges, respectively, so that the W bosons have $p = g = 0$.

The neutral weak nuclear force, briefly discussed in Section 5.2, is mediated by massive colorless Z^0 bosons, composed of both T and V rishons and $\bar{T}$ and $\bar{V}$ antirishons. This neutral weak nuclear force will be discussed in more detail in Chapter 11.

Chapter 7

Origin of Mass

7.1 Introduction

As discussed briefly in Section 4.2, in 1905 Einstein concluded from the theory of special relativity that the mass of a body m is a measure of its energy content E and is given by $m = E/c^2$, where c is the speed of light in a vacuum.

This relationship was first tested in 1932 by Cockcroft and Walton using the nuclear reaction:

$$p + \text{Li}^7 \to 2\alpha + 17.2 \text{ MeV}. \tag{7.1}$$

Cockcroft and Walton found that the decrease in mass in this disintegration process was consistent with the observed release of energy, according to $E = mc^2$. In 2005 this relationship has been verified [32] to within 0.00004%, using very accurate measurements of the atomic-mass difference, Δm, and the corresponding γ-ray wavelength to determine E, the nuclear binding energy, for isotopes of silicon and sulfur.

In the following Sections 7.2 and 7.3 the origin of mass in both the SM and the GM will be discussed. Section 7.4 provides a qualitative description of the mass hierarchy of all three generations of leptons and quarks, based upon the nature of the chromodynamic forces binding the rishons and/or antirishons constituents of the leptons and quarks, within the framework of the GM.

7.2 Origin of Mass in the SM

It has been emphasized [20] by Frank Wilczek (1951-) that approximate QCD calculations obtain the observed masses of the neutron and proton to an accuracy of within 10%. In these simplified calculations, the assumed constituents, quarks and gluons, are taken to be massless. Wilczek concludes that the calculated masses of the hadrons arise from both the energy stored in the motion of the quarks and the energy of the gluon fields, according to $m = E/c^2$. It is possible that if the calculations could be done more accurately, taking into account the intrinsic masses of the quarks, the calculated masses of the neutron and proton would agree more nearly with observation. Thus the mass of a nucleon arises essentially from internal energy. This origin of mass is supported by similar approximate QCD calculations of the masses of other hadrons.

Wilzcek [20] has also discussed the underlying principles giving rise to the internal energy, hence the mass, of a hadron. The nature of the gluon color fields is such that they lead to a runaway growth of the fields surrounding an isolated color charge. In fact all this structure (via virtual gluons) implies that an isolated quark would have an infinite energy associated with it. This is the reason why isolated quarks are not seen. Nature requires these infinities to be essentially cancelled or at least made finite. It does this for hadrons in two ways: either by bringing an antiquark close to a quark (i.e. forming a meson) or by bringing three quarks, one of each color, together (i.e. forming a baryon) so that in each case the composite hadron is colorless. However, quantum mechanics prevents the quark and the antiquark of opposite colors or the three quarks of different colors from being placed exactly at the same place. This means that the color fields are not exactly cancelled, although sufficiently it seems to remove the infinities associated with isolated quarks. The distribution of the quark-antiquark pairs or the system of three quarks is described by quantum mechanical wave functions. Many different patterns, corresponding to the various hadrons, occur. Each pattern has a characteristic energy, because the color fields are not entirely cancelled and because the quarks are somewhat localized. This characteristic energy, E, gives the characteristic mass, via $m = E/c^2$, of the hadron.

The above picture, within the framework of the SM, provides an understanding of hadron masses as arising mainly from internal energies, associated with the strong color interactions. However, it gives no indication as to the nature of the mass of a lepton, which is not associated with the strong color interaction.

In the SM, the masses of the elementary particles, the leptons, the quarks, and the W and Z particles, are interpreted [22] in a completely different way, arising from the existence of a 'condensate', analogous to the

Cooper pairs in a superconducting material. This condensate, called the Higgs field, was introduced by Brout, Englert and Higgs in 1964 in order to spontaneously break the $U(1) \times SU(2)_L$ local gauge symmetry of the electroweak interaction in the SM (see Chapter 5) to generate the masses of the W and Z gauge bosons. The Higgs field was also able to cure the associated fermion mass problem: the finite masses of the leptons and quarks cause the SM Lagrangian describing the system to violate the assumed local gauge invariance, since the CC weak nuclear interaction only acts upon the left-handed component of the fermion wavefunctions. The mass of each elementary particle of the SM is generated by the coupling of an originally massless particle to the Higgs field, which leads to the spontaneous symmetry breaking of the electroweak interaction and a corresponding mass term in the Lagrangian.

Unfortunately, the SM predicts only the relative masses of the W and Z bosons, while each fermion mass and that of the Higgs boson itself depend upon arbitrary coupling constants. Thus the introduction of a Higgs field within the framework of the SM leads to the introduction of fourteen new parameters. Indeed as pointed out by Holger Lyre (1965-) [33], the introduction of the Higgs field in the SM to spontaneously break the $U(1) \times SU(2)_L$ local gauge symmetry of the electroweak interaction, simply corresponds mathematically to putting in 'by hand' the masses of the elementary particles of the SM: the so-called Higgs mechanism does *not* provide any physical explanation for the origin of the masses of the leptons, quarks, and the W and Z bosons.

Thus to summarize: in the SM, most of the mass of ordinary matter (protons and neutrons) is attributed to the energy of their constituents, while the mass of each elementary particle (lepton, quark, or gauge boson) arises from the coupling of the particle to the Higgs field.

7.3   Origin of Mass in the GM

In the SM, the Higgs field was introduced to spontaneously break the $U(1) \times SU(2)_L$ local gauge symmetry of the electroweak interaction to generate the masses of the W and Z bosons, which were known to be massive, leading to the short-range nature of the weak interactions. This provided a consistent derivation of the so-called 'electroweak connection' [24], which gives a relation between the electromagnetic and the weak interactions in terms of two independent coupling constants, electric charge and weak charge. The Higgs field was also employed to generate the intrinsic masses of the leptons and quarks, although as pointed out by Lyre [33], this simply amounted mathematically to putting in 'by hand' the masses, without providing any physical explanation of their origin. In this way, the $U(1) \times SU(2)_L$ local gauge invariance, involving pure left-handed weak interactions was maintained.

In 2008 it was shown [25] that the 'electroweak connection' could be derived within the framework of the GM, which assumes that the weak interactions are *not* fundamental interactions, arising from a local gauge theory. Rather, the weak interactions are *residual* interactions of the strong nuclear force, responsible for binding the constituents of the leptons and quarks together. Thus, in the GM the weak interactions are 'effective' interactions and the mediators of these interactions may be massive without spoiling any local gauge invariance. The massive vector bosons, which mediate these effective weak interactions, are analogous to the massive mesons, which mediate the effective nuclear interactions between neutrons and protons. Consequently, the GM has no requirement for a Higgs field to generate the masses of the W and Z bosons.

In the GM, the elementary particles of the SM have a substructure, consisting of rishons and/or antirishons bound together by strong color interactions, mediated by massless hypergluons. This model is very similar to that of the SM in which the quarks are bound together by strong color interactions, mediated by massless gluons, to form hadrons. Since the mass of a hadron arises mainly from the energy of its constituents, the GM suggests that the mass of a lepton, quark or vector boson arises from a characteristc energy associated with its constituent rishons and/or antirishons and hypergluons. In particular, if the rishons, the antirishons and the hypergluons constituting the various leptons, quarks and vector bosons of the SM are all *massless*, then the masses of these particles arise entirely from the energy stored in the motion of the rishons and antirishons and the energy of the color hypergluon fields, E, according to Einstein's equation $m = E/c^2$. A corollary of this idea is: *if a particle has mass, then it is composite.*

In 1998 and 2001, two teams led by Takaaki Kajita (1959-) and Arthur McDonald (1943-), respectively, discovered independently using different techniques, neutrino oscillations, i.e. the phenomenon whereby a neutrino of one flavor, ν_e, ν_μ or ν_τ, is measured to have changed to a different flavor, e.g. $\nu_e \rightarrow \nu_\tau$ [23]. Kajita's team working at the Super-Kamiokande Observatory measured neutrino oscillations produced by cosmic ray neutrinos, while McDonald's team working at the Sudbury Neutrino Observatory measured that electron neutrinos (ν_e) produced by the Sun oscillated into both muon (ν_μ) and tau (ν_τ) neutrinos. These observations implied that neutrinos have mass, since according to quantum mechanics massless neutrinos cannot change their flavor, and consequently are expected to be composites, in agreement with the GM.

Thus to summarize: in the GM, the underlying mechanism for *all* mass has been postulated: the mass m of a body of ordinary matter is given by $m = E/c^2$, where E is the characteristic energy associated with the interactions, mediated by massless hypergluons, binding together the constituent massless rishons and/or antirishons of the body. Thus, unlike the SM, the GM provides a *unified* description of the origin of all mass, and has no requirement for a Higgs field to generate the mass of any particle. A corollary of this is: if a particle has mass, then it is composite.

7.4 Mass Hierarchy of Leptons and Quarks

Table 7.1 shows the observed masses of the charged leptons together with the estimated masses of the quarks: the masses of the neutral leptons have not yet been determined but are known to be very small. Although the mass of a single quark is a somewhat abstract idea, since quarks do not exist as particles independent of the environment around them, the masses of the quarks may be inferred from mass differences between hadrons of similar composition. The strong binding within hadrons complicates the issue to some extent but rough estimates of the quark masses have been made [3], which are sufficient for our purposes.

Charge	0	-1	$+\frac{2}{3}$	$-\frac{1}{3}$
Generation 1	ν_e	e^-	u	d
Mass	< 0.3 eV	0.511 MeV	5 MeV	10 MeV
Generation 2	ν_μ	μ^-	c	s
Mass	< 0.3 eV	106 MeV	1.3 GeV	200 MeV
Generation 3	ν_τ	τ^-	t	b
Mass	< 0.3 eV	1.78 GeV	175 GeV	4.5 GeV

Table 7.1. Masses of leptons and quarks.

The SM is unable to provide any understanding of either the existence of the three generations of leptons and quarks or their mass hierarchy indicated in Table 7.1. On the other hand, the GM suggests that both the existence and mass hierarchy of these three generations arise from the substructures of the leptons and quarks (see Chapter 6).

In the GM it is envisaged that the rishons and/or antirishons of each lepton or quark are very strongly localized, since to date there is no direct evidence for any substructure of these particles. Thus the constituents are expected to be distributed according to quantum mechanical wave functions, for which the product wave function is significant for only an *extremely small* volume of space so that the corresponding color fields are *almost cancelled*. The constituents of each lepton or quark are localized within a very small volume of space by strong color interactions acting between the colored rishons and/or antirishons. I call these *intrafermion* color interactions. However, between any two leptons and/or quarks there will be a residual interaction, arising from the color interactions acting between the constituents of one fermion and the constituents of the other fermion. I refer to these interactions as *interfermion* color interactions. These will be associated with the gravitational interaction and are discussed in the following Chapter 8.

The mass of each lepton or quark corresponds to a characteristic energy primarily associated with the intrafermion color interactions. It is expected that the mass of a composite particle will be greater if the degree of localization of its constituents is smaller (i.e. the constituents are on average more widely separated). This is a consequence of the nature of the strong color interactions, which are assumed to possess the property of 'asymptotic freedom' discovered by David Gross (1941-), Hugh Politzer (1949-) and Wilczek, whereby the color interactions become stronger for larger separations of the color charges. In addition, it should be noted that the repulsive electromagnetic interactions between charged T-rishons or between charged $\bar{T}$-antirishons will also cause the degree of localization of the constituents to be smaller causing an increase in mass.

There is some evidence for the above expectations. The electron consists of three $\bar{T}$-antirishons, while the electron neutrino consists of three neutral $\bar{V}$-antirishons. Neglecting the electric charge carried by the $\bar{T}$-antirishon, it is expected that the electron and its neutrino would have identical masses, arising from the similar intrafermion color interactions. However, it is anticipated that the electromagnetic interaction in the electron case will cause the $\bar{T}$-antirishons to be less localized than the $\bar{V}$-antirishons constituting the electron neutrino so that the electron will have a substantially greater characteristic energy and hence a greater mass than the electron neutrino, as observed. This large difference in the masses of the e^- and ν_e leptons (see Table 7.1) indicates that the mass of a particle is extremely sensitive to the degree of localization of its constituents.

Similarly, the up, charmed and top quarks, each containing two charged T-rishons, are expected to have a greater mass than their weak isospin partners, the down, strange and bottom quark, respectively, which contain only a single charged $\bar{T}$-antirishon. This is true provided one takes into account quark mixing in the case of the up and down quarks, although Table 7.1 indicates that the down quark is more massive than the up quark, leading to the neutron having a greater mass than the proton. This is understood within the framework of the GM since due to the manner in which quark masses are estimated, it is the *weak eigenstate* quarks, whose masses are given in Table 7.1. Since each succeeding generation is significantly more massive than the previous one, any mixing will noticeably increase the mass of a lower generation quark. Thus the weak eigenstate d'-quark, which contains about 5% of the mass eigenstate s-quark, is expected to be significantly more massive than the mass eigenstate d-quark. Consequently, the mass of a proton is expected to be smaller than the mass of a neutron and will be stable. I shall now discuss the mass hierarchy of the three generations of leptons and quarks in more detail.

It is envisaged that each lepton of the *first* generation exists in an antisymmetric three-particle colorless state, which physically assumes a quantum mechanical triangular distribution of the three differently colored identical rishons (or antirishons) since each of the three color interactions between pairs of rishons (or antirishons) is expected to be strongly attractive [30]. As indicated above, the charged leptons are predicted to have larger masses than the neutral leptons, since the electromagnetic interaction in the charged leptons will cause their constituent rishons (or antirishons) to be less localized than those constituting the uncharged leptons, leading to a substantially greater characteristic energy and a correspondingly greater mass.

In the GM, each quark of the *first* generation is a composite of a colored rishon and a colorless rishon-antirishon pair, $(T\bar{V})$ or a $(V\bar{T})$ (see Table 6.5). This color charge structure of the quarks is expected to lead to a quantum mechanical linear distribution of the constituent rishons and antirishons, corresponding to a considerably larger mass than that of the leptons, since the constituents of the quarks are less localized. This is a consequence of the character (i.e. attractive or repulsive) of the color interactions at small distances [30]. The general rules for small distances of separation are:

(1) Rishons (or antirishons) of like colors (or anticolors) repel: those having different colors (or anticolors) attract, unless their colors (or anticolors)

are interchanged and the two rishons (or antirishons) do not exist in an antisymmetric color state (e.g. as in the case of leptons);

(2) Rishons and antirishons of opposite colors attract but otherwise repel.

Furthermore, the electromagnetic interaction occurring within the up quark, leads one to expect it to have a larger mass than that of the d quark, although fortunately smaller than the d' quark.

Each lepton of the *second* generation is envisaged to basically exist in an antisymmetric three-particle color state, which physically assumes a quantum mechanical triangular distribution of the three differently colored identical rishons (or antirishons), as for the corresponding lepton of the first generation. The additional colorless rishon-antirishon pair, $(V\bar{V}^*)$ or $(V^*\bar{V})$, is expected to be attached externally to this triangular distribution, leading quantum mechanically to a less localized distribution of the constituent rishons and/or antirishons, so that the lepton has a significantly larger mass than its corresponding first generation lepton.

Each quark of the *second* generation has a similar structure to that of the corresponding quark of the first generation, with the additional colorless rishon-antirishon pair, $(V\bar{V}^*)$ or $(V^*\bar{V})$, attached quantum mechanically so that the whole rishon structure is essentially a linear distribution of the constituent rishons and antirishons. This structure is expected to be less localized, leading to a larger mass relative to that of the corresponding quark of the first generation, with the charmed quark having a greater mass than the strange quark, arising from the electromagnetic repulsion of its constituent two charged T-rishons.

Each lepton of the *third* generation is considered to basically exist in an antisymmetric three-particle color state, which physically assumes a quantum mechanical triangular distribution of the three differently colored identical rishons (or antirishons), as for the corresponding leptons of the first and second generations. The two additional colorless rishon-antirishon pairs, $(V\bar{V}^*)(V\bar{V}^*)$, $(V\bar{V}^*)(V^*\bar{V})$ or $(V^*\bar{V})(V^*\bar{V})$, are expected to be attached externally to this triangular distribution, leading to a considerably less localized quantum mechanical distribution of the constituent rishons and/or antirishons, so that the lepton has a significantly larger mass than its corresponding second generation lepton.

Each quark of the *third* generation has a similar structure to that of the first generation, with the additional two rishon-antirishon pairs $(V\bar{V}^*)$ and/or $(V^*\bar{V})$ attached quantum mechanically so that the whole rishon structure is essentially a linear distribution of the constituent rishons and antirishons. This structure is expected to be even less localized, leading to a larger mass relative to that of the corresponding quark of the second generation, with the top quark having a greater mass than the bottom quark, arising from the electromagnetic repulsion of its constituent two charged T-rishons.

The above is a qualitative description of the mass hierarchy of the three generations of leptons and quarks, based on the degree of localization of their constituent rishons and/or antirishons within the framework of the GM. However, in principle, it should be possible to calculate the actual masses of the leptons and quarks by carrying out QCD-type computations, analogous to those employed for determining the masses of the proton and other baryons within the framework of the SM.

Chapter 8

Gravity

8.1 Introduction

In this chapter I shall describe the path taken to create an alternative model to the SM of particle physics, which in 2002 I called the Generation Model (GM), since it was initially designed to describe the occurrence of three generations of the elementary particles of the SM. The current GM has been discussed in some detail in Chapter 6. However, it is of interest to provide a brief historical introduction to the development of the GM, which describes the occurrence of the three generations of leptons and quarks of the SM in terms of a composite model. This composite model also provides a solution of the enigma of gravity, mentioned in Chapter 1.

In Chapter 1 I also mentioned that in 2001 I attended a public lecture presented by Martinus Veltman concerning the "Facts and Mysteries in Elementary Particle Physics" in which it was emphasized that the *greatest puzzle of elementary particle physics was the occurrence of three families of elementary particles that have the same properties except for mass* in the SM.

It occurred to me that this three generation puzzle was analogous to the occurrence of similar patterns of the elements in Mendeleev's Periodic Table, in which groups of elements in columns of this table shared chemical properties and only differed in their atomic weights. All the atoms of these elements are not elementary but are *composites* of electrons, neutrons and protons.

This suggested to me, as it did to several other physicists much earlier, that the so-called elementary particles of the SM, the six leptons and the six quarks, as well as their antiparticles, were all actually composite particles. Furthermore, the equivalence in magnitude of the electric charges of the electron and proton, indicated that the electric charges of the quarks, comprising a proton, are intimately related to that of the electron, suggesting that the leptons and quarks are composed of the same kinds of building blocks. This in turn indicated that both leptons and quarks should be classified in terms of the same kinds of additive quantum numbers.

Consequently, the *nonunified* classification scheme of the SM (see Section 5.2) represented a major stumbling block for progress beyond the SM. In 2002 I discovered a new simpler and *unified* classification of the elementary leptons and quarks of the SM in terms of only three additive quantum numbers, charge (Q), particle number (p) and generation quantum number (g) (see Section 6.2). These three additive quantum numbers were *conserved* in all the interactions involving leptons and quarks (provided the particles mediating the interactions, γ, W, Z and gluons, have $p = g = 0$), unlike several of the quantum numbers , S, C, B and T, in the SM.

Thus my initial 2002 GM overcame the first major problem of the SM, involving a complicated nonunified classification scheme that presented a major stumbling block to the development of a composite model of the twelve elementary leptons and quarks of the SM.

Furthermore, as discussed in Section 6.2, my preference for all three additive quantum numbers to be conserved in all interactions led to an alternative quark-mixing procedure to that proposed by Cabibbo (see Section 5.2) to preserve the universalty of the CC weak nuclear interactions. This alternative quark-mixing procedure, placed the quark mixing within the quark states (wave functions), rather than in the CC weak nuclear interactions as proposed by Cabibbo. This different treatment of quark mixing in the 2002 GM had two important consequences.

First, it allowed a *conserved* additive quantum number, the generation quantum number, g, to be allotted to the physical quarks, and replaced the *non-conserved* S, C, B and T quantum numbers of the SM. Second, it required that hadrons be composed of CKM weak eigenstate quarks, rather than mass eigenstate quarks, as in the SM. This property of hadrons leads to further important consequences that I shall discuss later in Chapter 10.

The second major step in the development of an alternative model to the SM took place during the period 2003-2011 and resulted in the 2011 GM, which is a composite model of the leptons and quarks of the SM.

In the 2011 GM the basic building blocks of the first generation of leptons and quarks are assumed to be the two spin-$\frac{1}{2}$ *massless* rishons, the T-rishon with electric charge $Q = +\frac{1}{3}$ and the V-rishon with electric charge $Q = 0$, and their antiparticles, $\bar{T}$ and $\bar{V}$, introduced by Harari and Shupe in 1979. In the 2011 GM, each of these rishons carries a color charge, red, green or blue, analogous to those carried by quarks in the SM, while their antiparticles carry a color charge, antired, antigreen or antiblue, analogous to those carried by antiquarks in the SM.

In the 2011 GM, the higher generations of leptons and quarks employed a third kind of neutral rishon, called the U-rishon, and also its antiparticle the $\bar{U}$-antirishon, so that colorless neutral rishon-antirishon pairs could be added to the core substructures of the first generation of leptons and quarks, thereby maintaining essentially the same properties for the higher generations as those of the first generation.

In 2018 I found that the U-rishon could be considered to be a V-rishon in an excited state [27]. This is the current version of the GM and it has only the two building blocks, the T-rishon and the V-rishon, as well as their antiparticles, the $\bar{T}$-antirishon and the $\bar{V}$-antirishon.

8.2 Origin of Gravity

In the GM, both leptons and quarks have a substructure, consisting of spin-$\frac{1}{2}$ massless particles, rishons and/or antirishons, each of which carries a single color charge. These constituents of the leptons and quarks are bound together by strong color interactions, mediated by massless neutral vector hypergluons, acting between the color charged rishons and/or antirishons. These strong color interactions of the GM are analogous to the strong QCD color interactions of the SM, acting between color charged elementary quarks and/or antiquarks. In the GM, the strong chromodynamic force has been taken down one layer of complexity to describe the composite nature of leptons and quarks.

In the GM, the constituents of a lepton or quark must be very strongly localized, since to date there is no direct evidence for any substructure of these particles. Thus, the constituents are distributed according to quantum mechanical wave functions for which the product wave function, describing the composite particle is significant for only an extremely small volume of space.

In the case of the colorless leptons, this implies that the corresponding color fields are almost cancelled for distances outside the immediate vicinity of the lepton. In the case of the colored quarks, this means that the color charge of the quark is almost identical with that assumed in the SM, since the remaining color fields, corresponding to colorless rishon-antirishon pairs, are very nearly cancelled. Thus the dominant color interaction between quarks is essentially the same as that between rishons, so that the composite quarks of the GM behave very nearly like the elementary quarks of the SM.

In the GM, between any two colorless particles, electron, neutron or proton, there exists a residual interaction arising from the color interactions acting between the rishons and/or antirishons of one fermion and the color charged constituents of the other fermion. In 2009 I proposed that such *interfermion* color interactions may be identified with the usual gravitational interaction.

The gravitational interaction acts between bodies of ordinary matter with mass, and the mass of such bodies is essentially the total mass of its constituent electrons, neutrons and protons. In the GM, each of these three particles is composite and colorless. Indeed, all these three particles are considered to exist in a three-color antisymmetric state. A neutron or proton contains three quarks with one color charge of each of the three possible color charges: red, green and blue. An electron consists of three antirishons with one anticolor charge of each of the three possible anticolor charges: antired, antigreen and antiblue. The behavior of these particles with respect to the strong color interactions is expected basically to be the same, which suggests that the interfermion interactions of the GM between electrons, neutrons and protons have several properties associated with the usual gravitational interaction: universality, very weak strength and attraction.

First, the residual interaction between any two of the colorless particles, electron, neutron or proton, arising from the interfermion color interactions is predicted to be of a *universal* character.

Second, the residual interaction, mediated by massless neutral hypergluons, is expected to be *extremely weak*, since the constituents of each colorless particle are very strongly localized so that the strong color fields are almost completely cancelled at separations larger than the immediate vicinity of either of the two colorless particles. Consequently, the hypergluon self-interactions, resulting from their color charges, are also practically negligible and one may consider the color interactions using a perturbation approach.

Third, using the color factors [30] appropriate for the $SU(3)$ local gauge field, one finds that the residual color interactions between any two colorless particles, electron, neutron or proton, are each *attractive*.

Since the mass of a body of ordinary matter is essentially the total mass of its constituent electrons, neutrons and protons, the total interaction between two bodies of masses, m_1 and m_2, will be the sum of all the two particle contributions so that the total interaction will be proportional to the product of these two masses, $m_1 m_2$, provided that each two-particle interaction contribution is also proportional to the product of the masses of the two particles.

This latter requirement may be understood, if each electron, neutron or proton is physically considered to be essentially a quantum mechanical triangular distribution of three differently colored rishons or antirishons. In this case, each particle may be viewed as a distribution of three color charges throughout a small volume of space with each color charge having a certain probability of being at a particular point, determined by its corresponding color wave function. The total residual interaction between two colorless particles will then be the sum of all the intrinsic interactions acting between a particular triangular distribution of one particle with that of the other particle.

Now the mass m of each colorless particle is considered to be given by $m = E/c^2$, where E is a characteristic energy, determined by the degree of localization of its constituent rishons and/or antirishons. Thus the significant volume of space occupied by the triangular distribution of the three differently colored rishons or antirishons is larger the greater the mass of the particle. Moreover, due to *antiscreening effects*, known alternatively as a consequence of *asymptotic freedom*, which resulted from the theoretical work by Politzer, Gross and Wilczek and convinced the particle physics community of the validity of QCD as the theory of strong color interactions, the average strength of the color charge within each unit volume of the larger localized volume of space will be effectively increased. If one assumes that the mass of a particle is proportional to the integrated sum of the interfermion interactions within the significant volume of space occupied by the triangular distribution, then the total residual interaction between two such colorless particles will be proportional to the product of their masses.

Thus the residual color interaction between two colorless bodies of masses, m_1 and m_2, is proportional to the product of these masses, $m_1 m_2$, and moreover, neglecting the self-interactions of the hypergluons mediating the interaction, is expected to depend on the inverse square of their distance of separation r, i.e. as $1/r^2$, in accordance with Newton's universal law of gravitation.

However, the gravitational interaction in the GM, which I have identified with the interfermion color interactions between the colorless particles, electrons, neutrons and protons, has two additional properties arising from the self-interactions of the hypergluons, mediating the residual interaction: (1) asymptotic freedom and (2) color confinement. These two properties are absent for the electromagnetic interaction, since there are no corresponding photon self-interactions.

In the following Sections 8.3 and 8.4, I shall indicate how these two additional properties of the interfermion color interactions provide an understanding of both dark matter and dark energy.

Thus the solution to the enigma of gravity as indicated in Chapter 1 is the gravitational interaction in the GM as discussed above. In the GM, the usual gravitational interaction has been identified [5] with the residual interfermion color interactions acting between the colorless ordinary matter particles, electrons, neutrons and protons, which make up most of the ordinary mass of the Universe.

8.3   Dark Matter

The notion of 'dark matter' emerged from observations of large astronomical objects such as galaxies and clusters of galaxies as discussed in Section 4.4, which displayed gravitational effects that could not be accounted for by the visible matter: stars, hydrogen gas, etc., assuming the validity of Newton's universal law of gravitation.

It was concluded that such observations could only be described satisfactorily, if there existed stronger gravitational fields than those provided by the visible matter and Newtonian gravity. Such gravitational fields required either more mass or an appropriate modification of Newton's universal law of gravitation.

The conclusive observation from the rotation curves of spiral galaxies (see Section 4.4) was that there existed a major 'mass discrepancy' that was greater if larger distance scales were involved. This implied that if Newton's universal law of gravitation was approximately valid, as in the Solar System, considerably more mass was required to be present in each galaxy. This invisible matter was called *dark matter*. In 1974, Ostriker, Peebles and Yahil proposed that the rotation curves of spiral galaxies could most plausibly be understood if the spiral galaxy was embedded in a giant spherical halo of invisible dark matter that provided a large contribution to the gravitational field at large distances from the center of the galaxy. The only alternative to dark matter seemed to be a significant modification of Newton's universal law of gravitation to provide the required stronger gravitational field at larger distance scales. However, at that time, such a modification of Newtonian gravity was not considered a viable alternative.

The hypothesis of a dark matter spherical halo surrounding a spiral galaxy to account for the observed flat rotation curve of the galaxy has yet to be verified. One of the main difficulties is that the nature of the proposed dark matter is *unknown*.

Initially, massive compact halo objects (MACHOs), which are bodies having significant ordinary matter and are compact-like brown dwarfs, were searched for within the outer regions of galaxies. The conclusion from these observations was that at most 20% of a galactic halo consists of MACHOs, and the rest of the halo consists of nonbaryonic matter.

The only other known candidates for dark matter are the three neutrinos of the SM and the GM. However, it was demonstrated in 1983 by Carlos Frenk (1951-) and collaborators, that if dark matter consisted entirely of neutrinos, the large scale structure of the Universe would *differ* significantly from the observed one, since neutrinos are relativistic particles leading to a smooth large-scale structure.

The existence of dark matter in the universe suggests that one requires new physics beyond the SM. Three such particles have been searched for without success: (1) axions, (2) weakly interacting massive particles (WIMPS) and (3) sterile neutrinos. These three particles are all *hypothetical* particles, some of which have been introduced into particle physics in order to resolve certain perceived problems.

The axion was postulated in 1977 by Roberto Peccei (1942-2020) and Helen Arnold Quinn (1943-) in an attempt to resolve the strong CP problem in QCD by introducing a new pseudoscalar particle. To date various experiments have been carried out but *none* have successfully identified an axion particle.

A WIMP is considered to be a new elementary particle that only interacts via gravity and any other weak force. The basic goal of a direct detection of a WIMP is to measure the energy deposited when it interacts with nuclei in a detector, transferring energy to nuclei. Such direct detection experiments need to be carried out deep underground to prevent them being swamped by unwanted noise from cosmic ray particles.

The most favored WIMP is the lightest neutral stable particle, the *neutralino*, predicted by the supersymmetric (SUSY) theory of particle physics, which provides a significant relationship between elementary bosons and fermions. However, to date no evidence for any SUSY particle has been found either at the Large Hadron Collider (LHC) in CERN or in the many underground detection laboratories. At the LHC no previously *unknown* particles, which may be evidence of SUSY have been observed since the claimed detection of the Higgs boson, so that SUSY probably does not exist. In addition no WIMP has clearly been detected over several decades at any of the underground laboratories such as the large underground xenon (LUX) experiment in the Homestake Mine, Dakota.

However, there has been one claim of direct detection of dark matter from the DAMA-LIBRA experiment at the Gran Sasso laboratory. This experiment has observed a possible dark matter event rate that modulates annually as the Earth travels around the Sun, while the Solar System moves within the disk of the Milky Way and hence through the hypothesized galactic dark matter halo. The count rate is expected to depend upon the relative velocity of the detector and undergoes a modulation that peaks in June, when the relative velocity is at a maximum.

This observation of the DAMA-LIBRA experiment is controversial, since it has been excluded by observations from several direct detection experiments, including perhaps the most sensitive one, the LUX experiment.

Sterile neutrinos are also hypothetical neutral particles that emerged from the development of the electroweak theory by Glashow [25], who separated the neutrinos into left-handed and right-handed particles. The left-handed neutrinos interact via the left-handed weak interaction but the right-handed neutrinos do not, and only interact via gravity. The right-handed neutrinos correspond to the so-called sterile neutrinos.

The possible existence of sterile neutrinos arose in the development of the SM at a time when the neutrinos were considered to be *massless*. This is no longer the case so that the three 'normal' neutrinos in the SM are expected to have right-handed components with the same mass as the left-handed components and hence are unsuitable as candidates for dark matter. Moreover, the GM (see Section 6.3 and Chapter 11) indicates that such right-handed neutrinos do not exist.

The flat extended rotation curves (see Section 4.4) have provided the most convincing evidence for dark matter on galactic scales: in the flat part of a rotation curve, the rotation velocity is r independent, so that the centripetal acceleration goes as $1/r$ and the gravitational field must be decreasing as $1/r$. According to Poisson's equation of Newtonian theory, such a gravitational field (assuming spherical symmetry) must be sourced by a mass distribution with a r^{-2} profile. Since the visible mass density in the inner disk drops much faster (exponentially) than this, it is consistent to assume that the total mass distribution in the outer parts of the galaxy is approximately spherical. The conclusion is reached that each spiral galaxy must be immersed in a spherical dark matter halo with a mass density profile tending at large r, to r^{-2}. In Newtonian dynamics the circular velocity is expected to be

$$v(r) = \sqrt{GM(r)/r}, \tag{8.1}$$

where G is Newton's gravitational constant, $M(r) = 4\pi \int \rho(r)r^2 dr$ and $\rho(r)$ is the mass density profile, which should be falling $\propto 1/\sqrt{r}$ beyond the visible disk. The empirical fact that $v(r)$ is roughly constant implies the existence of a dark matter halo with $M(r) \propto r$ and $\rho \propto r^{-2}$.

However, the hypothesis of a dark matter halo surrounding a spiral galaxy to account for the observed flat rotation curve of a galaxy has yet to be verified and also there are several outstanding puzzles. First, the nature of the proposed dark matter is unknown, although it is now considered to be non-baryonic matter. Second, a dark matter halo has yet to be detected directly, although many searches have been carried out. Third, the density profile of a typical dark matter halo, within Newtonian physics, is required to be fine-tuned in order to produce the observed flat rotation curve of a spiral galaxy. Fourth, the lack of dark matter in globular clusters is still a mystery: large globular clusters have about the same mass as the smallest dwarf galaxies, although their diameter is only about a tenth the diameter of a typical dwarf galaxy, which are considered to have considerable amounts of dark matter.

Consequently, in view of these considerable uncertainties concerning the existence and nature of the proposed dark matter, there have been several attempts to modify Newton's law of gravitation instead of introducing dark matter.

The most successful of these early modifications is that published in 1983 by Mordehai Milgrom (1946-), known as the MOND (Modified Newtonian Dynamics) theory, as a possible alternative to dark matter. This theory is based upon describing two astronomical observations: (1) the flat rotational curves of spiral galaxies at large distances from their centers and (2) the Tully-Fisher empirical relation, named after Richard Tully (1943-) and James Fisher (1943-), which states that the intrinsic luminosity L (proportional to the total visible mass) of a spiral galaxy and the velocity, v_f, of the matter circulating at the extremities of the galactic disks are given by:

$$L \propto v_f^\alpha, \tag{8.2}$$

where α is approximately 4. In order to describe both the flat rotation curves of spiral galaxies and the Tully-Fisher relation, Milgrom suggested that gravity varies from the prediction of Newtonian dynamics for *low* accelerations. In particular, the transition from $1/r^2$ to $1/r$ gravity should occur below a 'critical acceleration' a_0 rather than beyond a critical distance r_0: the former leads to the Tully-Fisher relation, while the latter leads to:

$$L \propto v_f^2, \tag{8.3}$$

in gross disagreement with the Tully-Fisher relation. The modified law of gravity in terms of a_0 is:

$$g = GM/r^2 + (GMa_0)^{\frac{1}{2}}/r, \tag{8.4}$$

where g is the gravitational acceleration. The first term corresponds to distances for which the acceleration is much $> a_0$ and the second term corresponds to distances associated with the flat rotation curves, i.e. with v_f. Indeed the second term gives:

$$v_f = (GMa_0)^{\frac{1}{4}}, \tag{8.5}$$

which, if the mass to luminosity ratio, M/L, is roughly constant for galaxies, leads to the Tully-Fisher relation.

To summarize: MOND is an empirical modification of Newton's gravitational interaction that is designed to provide agreement with two overarching observational facts: (1) the flat rotation curves of spiral galaxies and (2) the Tully-Fisher relation. It achieves this aim by causing the gravitational interaction to change from $1/r^2$ for small distances, r to $1/r$ at large galactic distances as the gravitational acceleration becomes less than a critical small acceleration $a_0 \approx 1.2 \times 10^{-10}$ ms^{-2}.

Thus Milgrom's MOND theory, based upon a critical acceleration, describes the above two overarching facts emphasized by Jacob Bekenstein (1947-2015) as a requirement for any successful model, while a modified law of gravity, based upon a critical length scale does not. Indeed, Milgrom was the first to point out that any deviation from Newton's law in galactic systems had to appear below a critical acceleration in order to be consistent with observations.

In 1990, Robert Sanders [34] made perhaps the most interesting comments concerning both the MOND theory and the alternative hypothesis of dark matter.

First, in a conceptual sense, Sanders considers that it is better to view Milgrom's proposal as a modification of 'gravity' rather than as MOND suggests 'Newtonian dynamics', since the latter leads to problems such as the non-conservation of linear momentum of an isolated system.

Second, whether or not MOND is correct in its intended implication that gravity is non-Newtonian in the limit of low accelerations, it is certainly correct in the sense that mass discrepancies in galaxies tend to appear below a critical acceleration.

Third, even if a modification like MOND is in some sense correct, it is 'incomplete'. Such an idea must have its basis in a general theory of gravity, which makes some connection to more familiar physics. Viewed in this way, the central issue becomes the theory of gravity and not the explanation of flat rotation curves and the Tully-Fisher relation. These observations are simply signals that something is missing from the current theory of gravity. In particular, something is missing in the *general theory of relativity*, even in the classical limit, just as anomalous planetary precession indicated an incompleteness of Newtonian theory. If one takes this point of view, then the problem with the theory of general relativity is that it possesses the wrong weak field limit.

Fourth, Sanders concludes: "In the context of dark matter, any observed rotation curve can be explained by adjusting the parameters of some model for a dark halo or dark disk; the dark matter hypothesis can never be falsified. In this sense it is quite accurate to compare the dark matter hypothesis to the medieval system of crystal spheres. Any improvement in the accuracy of observations of the apparent motion of planets across the sky, could be accommodated in the system of unseen spheres simply by adding additional spheres. The system worked perfectly; it finally just became rather cumbersome. MOND, on the other hand, is quite analogous to Kepler's laws of planetary motion. Kepler's three rules worked no better than the older epicyclic hypothesis, but they were a far more efficient summary of the phenomena. Carrying the analogy further, it should be stressed that Kepler's rules were a mathematical prescription without physical content in the modern sense. It remained for Newton to unify such diverse phenomena as planetary motion and falling objects on earth in a single law of universal attraction. In a similar way, MOND, if it is correct, remains incomplete."

In Section 8.2, I have discussed the gravitational interaction in the GM: this interaction has been identified with the very weak, universal and attractive residual interfermion color interactions acting between the colorless particles, electrons, neutrons and protons, that essentially constitute the total mass of a body of ordinary matter. This interaction suggests a universal law of gravitation, which closely resembles Newton's original law that a spherical body of mass m_1 attracts another spherical body of mass m_2 by an interaction proportional to the product of the two masses and inversely proportional to the square of the distance, r, between the centers of mass of the two bodies:

$$F = H(r)m_1m_2/r^2 \,, \tag{8.6}$$

where Newton's gravitational constant is replaced by a function of r, $H(r)$.

This difference arises from the self-interactions of the hypergluons mediating the interfermion color interactions. These self-interactions cause antiscreening effects, which lead to an increase in the field strength of the residual color interaction acting between the two masses, so that H becomes an increasing function of r.

It is known from particle physics that the strong color fields tend to increase with the separation of color charges in order to confine quarks within baryons, so that one expects $H(r)$ to increase as a function of r. The flat rotation curves observed for spiral galaxies indicate that $H(r)$ is essentially a linear function of r. In this case, the modified law of gravity based upon gravity being identified with the very weak residual color interactions may be written:

$$g = GM/r^2 + GM\epsilon/(rr_S) \,, \tag{8.7}$$

where we have used:

$$H(r) = G(1 + \epsilon r/r_S) \,. \tag{8.8}$$

Here G is Newton's gravitational constant, ϵ is a factor representing the relative strengths of the modified and Newtonian gravitational fields and r_S is a radial length scale, dependent upon the radial mass distribution of the spiral galaxy, i.e. r_S varies from galaxy to galaxy.

In a spiral galaxy the gravitational interaction of a point mass at a distance r from the center of the galaxy will depend upon two factors: (i) the total mass M distributed within the sphere of radius r and (ii) the nature of the function $H(r)$. Thus for small values of r, these two factors will be entwined, each making a contribution to the orbital velocity of the point mass. However, for large values of r, only the second factor, $H(r)$, will make a significant contribution to the orbital velocity.

It should be noted that the above modification of the gravitational interaction will also cause a contribution to the advance of the perihelion of each planet of the Solar System, resulting from the second term in Eq. (8.7) (see Section 3.2).

As indicated above, the GM expression, Eq. (8.7), for the modified gravitational interaction is also associated with a radial parameter, r_S, which varies from galaxy to galaxy. Indeed, one can relate the modified terms in the gravitational acceleration expressions to obtain:

$$a_0 = GM\epsilon^2/r_S^2. \tag{8.9}$$

Thus the scale factor r_S may be regarded as the radial parameter beyond which the acceleration takes the value a_0 or less, and the value of r_S will depend upon the radial mass distribution of the galaxy. Equation (8.9) implies that the physical basis of the critical weak acceleration a_0 of the MOND theory is the existence of a radial parameter r_S, which defines a region beyond which the gravitational field behaves essentially as $1/r$. This occurs in the GM as a consequence of the universal nature of the weak color residual interaction identified as the universal gravitational interaction.

Equation (8.7) describes both the flat velocity rotation curves of spiral galaxies and also the correlation of their asymptotic orbit velocity, v_f, with their luminosity, the Tully-Fisher relation.

If one assumes that r_S is sufficiently large so that practically all the mass of a spiral galaxy is contained within a sphere of radius r_S (spherical symmetry is assumed for simplicity), then the asymptotic flat rotational velocity of a point mass for $r > r_S$ is given by

$$v_f = (GM\epsilon/r_S)^{\frac{1}{2}}. \tag{8.10}$$

Furthermore, the relative luminosity L may be written in terms of the average surface brightness Σ and the radial parameter r_S:

$$L = 4\pi\Sigma r_S^2, \tag{8.11}$$

so that

$$v_f^4 = (GM\epsilon/r_S)^2 \propto (M/L)^2(L/r_S)^2 \propto \tau^2\Sigma L, \tag{8.12}$$

where $\tau = M/L$. Equation (8.12) is the Tully-Fisher relation provided $\tau^2\Sigma$ is approximately constant, as has been found to be the case for spiral galaxies of both high and low surface brightness.

To summarize: the nature of the gravitational interaction in the GM leads to a modified law of gravity given by Eq. (8.7), which defines a scale factor r_S beyond which weak acceleration takes place and the gravitational field behaves essentially as $1/r$. Equation (8.7) describes both the flat rotation curves of spiral galaxies and also the correlation of their asymptotic orbital velocity with their luminosity (the Tully-Fisher relation). This modification of Newton's universal law of gravitation is very similar to that used in Milgrom's MOND theory, with the GM modification providing a physical basis as required by Sanders.

The continuing success of MOND theory in describing galactic mass discrepancy problems constitutes a strong argument against the existence of undetected dark matter haloes, consisting of unknown nonbaryonic matter, surrounding spiral galaxies.

However, a direct empirical proof of the existence of dark matter is claimed to be provided by two colliding galaxies known as the 'bullet cluster' [35]. Observations of the bullet cluster indicate that during the merging process, the dark matter, deduced from gravitational lensing, has passed through the collision point, while the baryonic component of matter, deduced from X-ray emission, has slowed down due to friction and has coalesed within a central region of the combined cluster. This separation of the two kinds of matter is claimed to provide evidence for dark matter. Unfortunately, a similar separation of the regions of non-Newtonian gravity in both the MOND and GM gravity theories is expected to occur in the merging of the bullet cluster.

8.4 Dark Energy

As discussed in Section 4.4, the notion of 'dark energy' arose from observations carried out in the later 1990s by two independent teams of astronomers that suggested that the expansion of the Universe is *accelerating*. These observations were very surprising and unexpected, since it was generally considered that the spatial expansion of the Universe should be slowing down due to the gravitational attraction of the galaxies.

The two teams analyzed supernovae of Type Ia, which are considered to be excellent standard candles across cosmological distances and allow the expansion history of the Universe to be measured by considering the relationship between the distance to an object and its redshift, which indicates how fast the supernova is receding from the Earth, if one assumes that the redshift arises entirely from a Doppler effect.

Both teams found that the supernovae observed about halfway across the observable Universe (6-7 billion light-years away) were dimmer than expected, assuming the existence of dark matter and Newton's universal law of gravitation, and concluded that the expansion of the Universe was accelerating rather than slowing down as expected.

The conclusion from this observation was that the Universe had to contain enough energy to overcome gravity. This energy was named *dark energy*. The amount of dark energy in the Universe was estimated [1] to be about 68% of the total mass-energy existing in the Universe.

Dark energy is a hypothetical form of energy, which is assumed to pervade the whole of space and causes the expansion of the Universe to accelerate at large cosmological distances. However, currently there exists no accepted physical theory of dark energy.

As indicated in Section 4.4, the SMC claims that dark matter and dark energy constitute about 27% and 68%, respectively, of the mass-energy content of the Universe. However, both dark energy and dark matter are simply names describing unknown entities, and consequently constitute two very dubious assumptions of the SMC.

In the previous Section 8.3 I have shown that the gravitational interaction in the GM provides an understanding of both the flat rotation curves of spiral galaxies and the Tully-Fisher relation without any requirement for the existence of dark matter. As I have indicated, the main effect arising from the self-interactions of the hypergluons, mediating the residual color interaction corresponding to the gravitational interaction of the GM, is to modify Newton's universal law of gravitation so that there is additional gravity at large galactic distances than that predicted by Newtonian mechanics.

However, the self-interactions of the hypergluons are expected to *cease* at a sufficiently large distance. This finite range arises from the *color confinement* property [4], which causes the residual color gravitational interaction to produce colorless particles, rather than continuing to modify Newton's universal law of gravitation by producing additional gravity at larger distances. This process takes place when the gravitational field energy is sufficient to produce the mass of a particle-antiparticle colorless pair. It is completely analogous to the 'hadronization process', involving the formation of hadrons out of quarks and gluons, which leads to the finite range ($\approx 10^{-15}$ m) of the strong color interaction in the SM.

In the gravitational case, the relative intrinsic strength of the interaction is about 10^{-41} times weaker than the strong color interaction at 10^{-15} m [4]. This suggests that the equivalent process to hadronization in the gravitational case should occur at a cosmological distance of about 10^{26} m, i.e. roughly 10 billion light-years (one light-year is $\approx 10^{16}$ m). This result agrees well with the observations of distant Type Ia supernovae, which indicate that the onset of the accelerating expansion of the Universe occurs at about 6 billion light-years.

The above finite range of the gravitational field, estimated to be roughly 10 billion light years, agrees very well with the 1997-1998 observations of Perlmutter, Riess, Schmidt and their collaborators of Type Ia supernovae about half-way across the observable Universe (6-7 billion light years), which they found to be *dimmer* than expected.

Unfortunately, both the estimated finite-range effect of the gravitational force and the Type Ia supernovae observations are subject to large errors. In particular, the supernovae observations of 1998-1999 suffer from (1) the small number of data points; (2) the assumption that the redshift arises entirely from a pure Doppler effect, i.e. there is no contribution from the 'tired-light' effect; and (3) the assumption that the increase in the gravitational field at large galactic distances is described by dark matter haloes and Newton's universal law of gravitation.

However, I consider that the supernovae observations did discover an *anomalous effect* of the SMC that corresponds to the finite-range effect predicted by the GM.

Hopefully, this may be reconciled when much larger data observations become available in the early 2020s, so that both dark matter and dark energy are recognized as consequences of the GM quantum theory of gravity: dark matter may be replaced by the modification of the Newtonian gravitational field at large galactic distances, arising from the *antiscreening effects*, and dark energy may be replaced by the finite-range of the gravitational field, arising from the *color confinement* property of the quantum gravitational force in the GM.

8.5   Gravity and the Photon

In 2016 I attended the 19th Workshop on *"What Comes Beyond the Standard Models"* held in Bled, Slovenia: the Standard Models referred to in the title are the SM of particle physics and the SMC, both of which were of considerable interest to me.

During the discussion of my work on gravity within the framework of the GM, the subject was raised: What property of the photon causes the gravitational deflection of light rays by massive bodies like the Sun?

This property of light rays to be deflected by the Sun has been discussed briefly in Section 3.2. Indeed, in 1919 Eddington and collaborators observed the deflection of light rays from distant stars passing near the Sun during a total solar eclipse, in order to differentiate between Einstein's general theory of relativity and Newton's theory of gravity.

The exact solution of Einstein's equation of general relativity, found by Schwarzschild in 1916 for the curved spacetime surrounding a spherical massive object, predicted that the deflection of light rays was about 1.76 arc-seconds. This value is exactly *twice* the value predicted by Newton's theory of gravity, if photons, i.e. quanta of energy, are subject to Newtonian gravitation in the same manner as massive particles.

Eddington, the leader of the 1919 expeditions, concluded that the observations favoured Einstein's theory rather than Newton's theory, although the observations were not sufficiently accurate to be conclusive. However, later measurements using microwaves have confirmed Eddington's conclusion.

After some contemplation, following the Bled Workshop, it occurred to me that the deflection of photons by the Sun could be described successfully, if the photon was assumed to be the standard *singlet state* of the corresponding QCD color octet of hypergluons binding together the rishons and antirishons of the leptons and quarks.

In the GM, the singlet state is a hypergluon, which is massless, electrically neutral, colorless and has spin-1 and $U(1)$ symmetry. In addition, this singlet state is expected to carry both a colorless set of three color charges and a colorless set of three anticolor charges. Furthermore, it interacts with ordinary matter via the new gravitational interaction of the GM, based upon residual color interactions. Consequently, the GM predicts that such photons will be deflected by massive bodies, such as the Sun, by an amount that is *twice* the value predicted by the new gravitational interaction of the GM for the deflection of ordinary matter.

Thus, the gravitational interaction of the GM, predicts the same deflection of light rays, consisting of photons, assumed to be singlet state hypergluons, in agreement with Einstein's theory of general relativity but without any warping of spacetime. It should also be noted that in principle, the gravitational interaction of the GM is also able to account for the anomalous advance of the perihelion of Mercury in terms of the additional term in the gravitational interaction.

Chapter 9

The Matter-Antimatter Asymmetry Problem

9.1 Introduction

The matter-antimatter asymmetry problem, corresponding to the virtual absence of antimatter in the Universe, is one of the greatest mysteries of cosmology.

The SMC [1] assumes that the Universe was created in the Big Bang from pure energy, and is now composed of about 5% ordinary matter, 27% dark matter and 68% dark energy. In Chapter 8, it was shown that both dark matter and dark energy can be replaced by a modified Newtonian-like universal gravitational field, so that the only matter in the Universe is ordinary matter, and all the remaining energy is within the gravitational and electromagnetic fields.

It is also generally assumed that the Big Bang produced *equal numbers* of particles and antiparticles. This leads to the matter-antimatter asymmetry problem, since the Universe today is considered to consist almost entirely of matter (particles) rather than antimatter (antiparticles): Where have all the antiparticles gone?

An understanding of the asymmetry problem requires both knowledge of the physical nature of the Big Bang and a precise definition of *matter*. Unfortunately, knowledge of the Big Bang is currently far from complete and matter has not been defined precisely within the framework of the SM.

The prevailing model of the Big Bang is based upon the general theory of relativity. According to this theory, extrapolation of the expansion of the Universe backwards in time yields an infinite density and temperature at a finite time, approximately 13.8 billion years ago. Thus the 'birth' of the Universe appears to be associated with a *singularity*. This indicates a breakdown of the general relativity theory but also all the laws of physics.

115

This is a serious impediment to understand the matter-antimatter (i.e. the particle-antiparticle) asymmetry problem. Consequently, this problem will be discussed in terms of the observed nature of the Universe, ignoring the singularity.

A consistent definition of the terms *matter* and *antimatter* is the following: *matter* is built of elementary *matter* particles and *antimatter* is built of elementary *antimatter* antiparticles.

In the SM the elementary *matter* particles are assumed to be the leptons and quarks so that electrons, neutrons and protons, which constitute the 5% ordinary matter in the Universe are all *matter*. In the GM, the elementary *matter* particles are rishons and the elementary *antimatter* antiparticles are antirishons, so that electrons, neutrons and protons are not all *matter*.

In this chapter, the various attempts to understand the matter-antimatter problem within the framework of the SM will be described. To date all these attempts have failed to provide an acceptable explanation of the matter-antimatter asymmetry so that the general conclusion is that physics beyond the SM is required for this purpose.

In my view, the main reason that the SM fails is that the SM assumes that the leptons and quarks are all elementary particles, so that the matter/antimatter nature of leptons and quarks may be decided by 'pure convention': in the SM both leptons and quarks are assumed to be matter particles.

In the following, it will be demonstrated that the GM, which considers all the leptons and quarks as composite particles, provides a possible solution to the matter-antimatter asymmetry problem.

9.2   Matter-Antimatter Asymmetry Problem and the SM

Within the framework of the SM, the matter-antimatter asymmetry problem is generally considered to be related to the baryon-antibaryon asymmetry problem, i.e. the imbalance of baryonic matter and antibaryonic matter in the observable Universe, which seems to consist almost entirely of hydrogen and helium atoms, rather than antihydrogen and antihelium atoms.

At first sight, this observation is rather surprising since the origin of the Universe in the Big Bang is generally considered to have produced equal numbers of baryons and antibaryons, which as the Universe cooled should have annihilated in pairs to pure energy, so that the Universe would be empty of matter.

In the SM this is assumed to imply that although the Universe was originally perfectly symmetric in baryons and antibaryons, during the cooling period some physical process contributed to a small imbalance in favor of baryons. This indicates that contrary to the current laws of physics, baryon number must be violated in some physical process.

This was proposed by Andrei Sakharov (1921-1989) in 1967. Sakharov proposed a set of three necessary conditions, within the framework of the SM, that a physical process must satisfy to produce baryons and antibaryons at different rates: (1) violation of baryon number; (2) violation of both charge conjugation symmetry, C, and charge conjugation-parity, CP; and (3) the process must not be in thermal equilibrium.

Violation of baryon number is required to produce an excess of baryons over antibaryons, while C symmetry violation ensures the non-existence of processes, which produce an equivalent excess of antibaryons over baryons. Similarly, violation of CP symmetry is required so that equal numbers of left-handed baryons and right-handed antibaryons as well as equal numbers of right-handed baryons and left-handed antibaryons are not produced. Finally, a departure from thermal equilibrium must play a role so that CPT symmetry does not ensure compensation between processes increasing and decreasing the baryon number [36].

The first Sakharov criterion: violation of baryon number, would be achieved if antiprotons or protons decayed into light subatomic particles such as a neutral pion and an electron or positron, respectively. However, there is currently *no* experimental evidence that such 'direct' violations of baryon number occur.

Thus researchers turned their attention to 'indirect' violations of baryon number, which are concerned with Sakharov's second criterion: CP violation, which indicates the possibility that some physical process may distinguish between matter and antimatter.

Both the electromagnetic and strong nuclear interactions are symmetric under C and P, and consequently they are also symmetric under the product CP. However, this is not necessarily the case for the CC weak nuclear interaction, which violates both C and P symmetries. Indeed, the 1964 discovery of the decay of the long-lived K_L^0 meson to two charged pions by James Cronin (1931-2016), Val Fitch (1923-2015) and collaborators, brought about the surprising conclusion that CP may also be violated in CC weak nuclear interactions. The violation of CP in CC weak nuclear interactions implies that such physical processes could lead to indirect violation of baryon number so that matter would be preferred over antimatter creation. However, later work (see Chapter 10) suggests that the two pion decay of the long-lived neutral kaon may be described in the GM without CP violation.

In the SM, any CP violation originates from the CC weak nuclear interactions that change both the charge and the flavor of quarks. The six known quarks (see Section 5.2) consist of three up-like quarks with charge $Q = +\frac{2}{3}$: up (u), charmed (c) and top (t); and three down-like quarks with charge $Q = -\frac{1}{3}$: down (d), strange (s) and bottom (b). The CC weak nuclear interactions, mediated by W^+ and W^- bosons, cause each up-like quark to turn into a down-like quark and vice-versa. The transition amplitudes for the nine combinations are given by the Cabibbo-Kobayashi-Maskawa (CKM) matrix elements:

$$V_{CKM} = \begin{pmatrix} V_{ud} & V_{us} & V_{ub} \\ V_{cd} & V_{cs} & V_{cb} \\ V_{td} & V_{ts} & V_{tb} \end{pmatrix} \approx \begin{pmatrix} 0.9742 & 0.2253 & 0.0035 \\ 0.2252 & 0.9734 & 0.0412 \\ 0.0087 & 0.0404 & 0.9991 \end{pmatrix}, \qquad (9.1)$$

where, e.g. V_{ud} is the transition amplitude and $|V_{ud}|^2$ is the transition probability for the CC weak nuclear interaction:

$$u \rightarrow d + W^+. \qquad (9.2)$$

Prior to the quark model of Gell-Mann and Zweig, in 1963 Cabibbo had introduced several matrix elements of the CKM matrix in order to preserve the universality of the CC weak nuclear interaction for the first and second generations of quarks (see Section 5.2). Indeed, in the V_{CKM} matrix the value of V_{ud} and V_{us} correspond to $\cos \theta_c$ and $\sin \theta_c$, respectively, where θ_c is the famous Cabibbo angle.

In 1973 Kobayashi and Maskawa proposed another method of introducing CP violation into the SM: by extending the idea of 'Cabibbo mixing' to three generations, they demonstrated that this allowed a complex phase to be introduced into the CKM matrix, permitting CP violation to be directly incorporated into the weak nuclear interaction. This proposal also predicted the existence of three additional quarks that were unknown at that time: the charmed quark, the bottom quark and the top quark, discovered in 1974, 1977 and 1995, respectively.

However, avoiding rather complex mathematical details, the Kobayashi-Maskawa CP violation introduced into the SM is found to be tiny, primarily because of the smallness of the relevant matrix elements. Consequently, any physical process that produces more matter than antimatter would have been ineffectual. Although the excess of matter over antimatter is generally considered to have been only one part in a billion, the effect of the Kobayashi-Maskawa CP violation process falls far short of even this very small amount by many orders of magnitude. Indeed it is estimated that the baryon excess produced by the Kobayashi-Maskawa CP violation process is only sufficient to provide the baryons of a single galaxy in the Universe, which comprises billions of galaxies.

The third criterion of Sakharov: departure from thermal equilibrium, is generally assumed to also occur within the electroweak sector of the SM during the so-called electroweak phase transition. This is assumed to be a first order transition between the state in which the W and Z gauge bosons are massless to a state in which they are massive. The massive W and Z bosons are assumed to arise as a result of some unknown mechanism which breaks the electroweak symmetry, and it is during this electroweak spontaneous symmetry breaking that departure of thermal equilibrium takes place.

Thus the SM does provide possible physical processes, which satisfy all three necessary criteria of Sakharov. However, the assumed physical processes do not seem capable of providing an acceptable explanation for the matter-antimatter asymmetry. The general conclusion is that physics beyond the SM is required for this purpose.

9.3 Matter-Antimatter Asymmetry Problem and the GM

In Section 9.2 I have discussed the matter-antimatter asymmetry problem within the framework of the SM. For many decades now the SM has been unable to provide an understanding of the asymmetry problem. The main reason I believe is that the SM assumes that the leptons and quarks are all elementary particles so that the matter/antimatter nature of leptons and quarks may be decided by pure convention. In the SM both leptons and quarks are assumed to be matter particles.

In the SM it is assumed that the Big Bang initially produces numerous elementary particle-antiparticle pairs such as electron-positron pairs and quark-antiquark pairs by converting energy into mass according to $m = E/c^2$. Thus the early Universe consisted of a soup of particle-antiparticle pairs continually being created and annihilated. Later, as the Universe cooled, the quarks and antiquarks would form protons, neutrons, antiprotons, antineutrons, etc., and eventually, together with electrons and positrons, atoms of hydrogen, antihydrogen, helium and antihelium. These would later annihilate in pairs until only atoms of hydrogen and helium prevailed.

However, it is unlikely that either electrons or positrons would prevail so that neither hydrogen atoms nor antihydrogen atoms would prevail, since both electrons and positrons are elementary particles in the SM and consequently the creation and annihilation of electron-positron pairs constitute a *unique* process, so that the Universe will always contain an equal number of electrons and positrons.

In the GM it is expected that the Big Bang initially produces numerous elementary rishon-antirishon pairs corresponding to the two kinds of rishons. Then as the Universe cooled, the rishons and antirishons would form leptons, quarks and their antiparticles and eventually atoms of hydrogen, antihydrogen, helium and antihelium. Later these would annihilate in pairs until only atoms of hydrogen and helium prevailed.

In order to understand the matter-antimatter asymmetry in the GM, it is necessary to define the matter/antimatter nature of composite particles. Historically, the term 'particle' defines matter that is naturally occurring, i.e. electrons, protons, hydrogen atoms, etc. This is consistent within the SM in which the electron and the up and down quarks are elementary particles and are defined conventionally as matter. However, it is inconsistent within the GM in which the electron and the up and down quarks are composite particles consisting of rishons and/or antirishons. In the GM, rishons are considered to be matter, while antirishons are considered to be antimatter, since rishon-antirishon pairs are considered to be created/annihilated in the Big Bang.

In the GM (see Table 6.4), the elementary rishons and antirishons have $p = +\frac{1}{3}$ and $p = -\frac{1}{3}$, respectively. Thus the particle number p allotted to a composite lepton or quark reflects its degree of matter or antimatter nature. In the GM the quarks are composed of both rishons and antirishons so that they have both a matter and an antimatter nature. On the other hand, an electron, consisting of three $\bar{T}$-antirishons has $p = -1$ and is pure antimatter.

The solution of the matter-antimatter asymmetry problem involves the particle number additive quantum number p of the GM: in particular the values of p corresponding to a weak eigenstate u-quark ($p = +\frac{1}{3}$), a weak eigenstate d'-quark ($p = +\frac{1}{3}$) and an electron ($p = -1$), since these are the constituents of atoms. The values of $p = \frac{1}{3}$ of the weak eigenstate quarks correspond to the values of their baryon number in the SM, while the value of $p = -1$ of the electron, corresponds to minus the value of the lepton number of the electron in the SM. In the GM, the electron consists entirely of antirishons, i.e. antiparticles, while in the SM it is assumed to be a particle, although as I have indicated earlier, there is no a priori reason for this assumption based solely upon convention. It should be noted that the matter/antimatter nature of an electron in the GM is not merely a revised definition of the term 'matter' but is a requirement for consistency of the nature of the constituents of the electron and the initial particle-antiparticle nature of the Universe in the Big Bang: the elementary particles in the SM, leptons and quarks, and in the GM, rishons, are *different*.

In the GM, the proton consists of three weak eigenstate quarks, two u-quarks and one d'-quark, so that the proton has particle number $p = +1$. Consequently, a hydrogen atom, consisting of one proton and one electron has particle number $p = 0$. The hydrogen atom in the GM consists of an equal number of rishons and antirishons, so that $p = 0$ and there is no asymmetry of matter and antimatter there.

In the GM, the neutron is composed of three weak eigenstate quarks, one u-quark and two d'-quarks, so that the neutron also has particle number $p = +1$. Consequently, a helium atom, consisting of two neutrons, two protons and two electrons has particle number $p = +2$: the helium atom in the GM consists of six more rishons than antirishons, i.e. more matter than antimatter. In the GM, during the formation of helium in the aftermath of the Big Bang, it is assumed that an equivalent surplus of antimatter was formed as neutrinos, which have $p = -1$, so that overall equal numbers of rishons and antirishons prevailed. This assumption is a consequence of the conservation of p in *all* interactions in the GM.

A further consequence of the conservation of p in all interactions is that, if the initial state of the Universe has $p = 0$ following the production of equal numbers of particles and antiparticles in the Big Bang, then the GM *predicts* that the present state of the Universe should also have $p = 0$.

To summarize: the ordinary matter present in the Universe has an overall particle number of $p = 0$, so that it contains equal numbers of both rishons and antirishons. This implies that the original antimatter created in the Big Bang is now contained within the stable composite leptons, i.e. electrons and neutrinos, and the stable composite quarks, i.e. the weak eigenstate u-quarks and d'-quarks, comprising the protons and neutrons. The hydrogen, helium and heavier atoms, following the fusion processes in stars, all consist of electrons, protons and neutrons. This explains where all the antiparticles have gone. Thus there is *no* matter-antimatter asymmetry in the present Universe. However, it does not explain why the Universe consists primarily of hydrogen atoms and not antihydrogen atoms. It is suggested that this hydrogen-antihydrogen asymmetry may be understood as follows.

In the GM, an antihydrogen atom consists of the same equal numbers of rishons and antirishons as a hydrogen atom, although the rishons and antirishons are differently arranged in the two systems. In particular, the three T-rishons of the positron and the three $\bar{T}$-antirishons of the antiproton, within the antihydrogen atom, are identical to the three T-rishons of the proton and the three $\bar{T}$-antirishons of the electron within the hydrogen atoms. This implies that both hydrogen atoms and antihydrogen atoms should be formed during the aftermath of the Big Bang with about the same probability. In fact, estimates from the CMB data suggest that for every billion hydrogen-antihydrogen pairs, there was just one extra hydrogen atom [4].

It is proposed that this extremely small difference, one extra hydrogen atom in 10^9 hydrogen-antihydrogen pairs, may arise from small statistical fluctuations associated with the complex many-body processes involved in the formation of either a hydrogen atom or an antihydrogen atom. The uniformity of the Universe, in particular, the lack of antihydrogen throughout the Universe, indicates that such statistical fluctuations took place prior to the 'inflationary period' [4] associated with the Big Bang scenario.

It should be noted that following the annihilation of the hydrogen-antihydrogen pairs that the Universe is expected to be dominated by photons and neutrinos, rather than matter particles. However, it is the relatively small number of matter particles that allows us to exist. It should also be noted that the SM does *not* predict this possibility to occur.

Chapter 10

Mixed-Quark States in Hadrons

10.1 Introduction

As discussed in Section 6.2, the GM postulates that hadrons are composed of weak eigenstate quarks rather than mass eigenstate quarks as in the SM. This gives rise to several important consequences.

First, the occurrence of mixed-quark states in hadrons implies the existence of higher generation quarks in hadrons. In particular, the GM predicts that the proton, having two u and one d' quarks contains about 1.7% of strange quarks, while the neutron, having one u and two d' quarks, contains about 3.4% of strange quarks, since the d' quark is given by the matrix elements of the CKM matrix in Eq. (9.1) as

$$d' = 0.9742d + 0.2253s, \tag{10.1}$$

so that $|V_{ud}|^2 \approx 0.949$ and $|V_{us}|^2 \approx 0.051$.

It is interesting to note that several experiments conducted in Mainz, Germany and at the Jefferson Laboratory, New Norfolk, USA in 2005 and more recently at the LHC in CERN in 2017, have provided some evidence for the existence of strange quarks in the proton. In the SM, such strange quarks form part of the 'sea' of quark-antiquark pairs arising from spontaneous pair creation from the gluons inside the proton. In the GM, one may have a combination of the sea quark-antiquark pairs arising from spontaneous pair creation from the hypergluons inside the proton, and in addition the strange quarks present inside the proton arising from the $\approx 5\%$ content of the strange quarks in the d'-quark within the proton. Unfortunately, none of the above rather difficult experiments have been able to determine sufficiently accurately the actual overall percentage of strange quarks within the proton, predicted by the GM to be approximately 1.7%, if one neglects the small contribution arising from spontaneous pair creation.

Second, the presence of strange quarks in nucleons provides an understanding of why the mass of a neutron is greater than the mass of a proton, so that the proton is stable. Although the mass of a single quark is a somewhat abstract idea, since quarks do not exist as particles independent of the environment around them, the masses of the quarks may be inferred from mass differences between hadrons of similar composition. The strong binding within hadrons complicates the issue to some extent but rough estimates of the quark masses have been made [3]: these are (u) 5 MeV, (d) 10 MeV, (c) 1.3 GeV, (s) 200 MeV, (t) 175 GeV and (b) 4.5 GeV. As expected (see Section 7.4) the top quark is much more massive than the bottom quark and also the charmed quark is considerably more massive than the strange quark. On the other hand, while the masses of the up and down quarks are more comparable, it is the down quark that is more massive than the up quark, leading to the neutron having a greater mass than the proton. The SM has no natural explanation for this but within the GM framework there exists a possible answer to this inconsistency: due to the manner in which masses are estimated in the GM, it is the *weak eigenstate* quarks $(d'$, s' and $b')$, whose masses are given above. Since each succeeding generation is significantly more massive than the previous one, any mixing will noticeably increase the mass of a lower generation quark. Thus the weak eigenstate d'-quark, which contains about 5% of the mass eigenstate s-quark, is expected to be significantly more massive than the mass eigenstate d-quark. This is a possible explanation for the 'down' quark having a larger mass than the up quark, contrary to the strange-charmed and bottom-top quark doublets.

Another consequence of the presence of mixed-quark states in hadrons is that mixed-quark states may have *mixed parity*. In the GM, the constituents of quarks are both rishons and antirishons. If one assumes the simple convention that all rishons have positive parity and all their antiparticles have negative parity, one finds that the down and strange quarks have *opposite* intrinsic parities, according to the proposed structures of these quarks in the GM: the d-quark (see Table 6.5) consists of two rishons and one antirishon so that the parity of the d-quark is $P_d = -1$, while the s-quark (see Table 6.6) consists of three rishons and two antirishons so that the parity of the s-quark is $P_s = +1$. The u-quark consists of two rishons and one antirishon so that $P_u = -1$, and the antiparticles of these three quarks have the corresponding opposite parities: $P_{\bar{d}} = +1$, $P_{\bar{s}} = -1$ and $P_{\bar{u}} = +1$.

The parity of charged pions has played a significant role in the overthrow of both parity (P) conservation (see Section 4.3) and combined charge-conjugation-parity (CP) conservation [37] in CC weak nuclear interactions. This will be discussed in the following Section 10.2.

10.2 Parity of Charged Pions and CP Violation

In the SM the parity of charged pions is assumed to be $P_\pi = -1$. This value was first obtained in 1954, within the theoretical framework of the time, by William Chinowsky (1929-) and Steinberger using the capture of negatively charged pions by deuterium to form two neutrons:

$$\pi^- + D \to n + n. \tag{10.2}$$

This experiment was carried out prior to the 1964 quark model of mesons and baryons (see Section 4.2), so that in the analysis of the experiment, the pion, the proton and the neutron were each assumed to be elementary particles with known spins 0, $\frac{1}{2}$ and $\frac{1}{2}$, respectively. In addition, it was also known that the pion is captured by the deuteron, having spin 1, from an S state so that the relative orbital angular momentum in the initial state is $l_i = 0$, and thus the total angular momentum of the initial state is $j_i = 1$. An underlying assumption in the analysis was that the pion had a definite intrinsic parity: $P_\pi = +1$ or $P_\pi = -1$.

Thus, since the intrinsic parity of the deuteron was known to be $P_D = +1$, if one assumes that the neutron and proton have the same intrinsic parity, the parity of the initial state is given by

$$(-1)^{l_i} P_\pi P_D = (-1)^0 P_\pi P_D = P_\pi. \tag{10.3}$$

The parity of the final state for relative orbital angular momentum l_f is

$$P_n P_n (-1)^{l_f} = (-1)^{l_f}, \tag{10.4}$$

assuming that the neutron has a definite intrinsic parity $P_n = +1$ or $P_n = -1$. Consequently, if parity is conserved in the strong nuclear interaction:

$$P_\pi = (-1)^{l_f}. \tag{10.5}$$

Since the neutrons are identical fermions, the allowed antisymmetric states $\left({}^{2S+1}L_J\right)$ of the two neutrons are ${}^1S_0$, ${}^3P_{0,1,2}$, ${}^1D_2$, ${}^3F_{2,3,4}$, etc. The only final state with $j_f = j_i = 1$ is the ${}^3P_1$ state, so that $l_f = 1$ and hence $P_\pi = -1$.

Following the adoption of the 1964 quark model as part of the SM, the parity of charged pions remained accepted as $P_\pi = -1$: in the quark model, the π^- was proposed to be a combination of a down quark (d) and an up antiquark $(\bar{u})$, i.e. $\pi^- \equiv [d\bar{u}]$, so assuming that elementary quarks have $P_q = +1$, while their elementary antiquarks have $P_{\bar{q}} = -1$, the charged pions have $P_\pi = -1$, in agreement with the result of Chinowsky and Steinberger.

This value of the parity of charged pions, $P_\pi = -1$, led to the overthrow of both parity conservation in 1957 and combined charge-conjugation-parity (CP) conservation in 1964 in CC weak nuclear interactions. I shall now describe briefly how this came about.

As discussed in Section 4.3, the discovery during 1947-1953 of two new particles (see relations (4.15) and (4.16)), the θ^+ and the τ^+, which decayed into two and three pions, respectively, via a CC weak nuclear interaction, and were found to be *indisinguishable* apart from their decay mode, since their charges, masses and lifetimes were found to be essentially the same, presented a problem called the *theta-tau puzzle*. This led Lee and Yang to question whether parity was conserved in CC weak nuclear interactions, and this question was quickly answered in the negative: parity conservation was violated in CC weak nuclear interactions.

The first experiment to investigate parity conservation in CC weak nuclear interactions was carried out by Wu and collaborators in 1956 employing the β-decay of polarized Co^{60} nuclei:

$$Co^{60} \rightarrow Ni^{60} + e^- + \bar{\nu}_e, \qquad (10.6)$$

and noting the direction of emission of the electrons with respect to the direction of the spin of the Ni^{60} nucleus. If parity was conserved, it was anticipated that an equal number of electrons would be emitted both parallel and antiparallel to the spin of the Ni^{60} nucleus. The final result of this experiment was that many more electrons were emitted in the antiparallel direction than in the parallel direction, so that parity symmetry was *violated*. This experiment is described in more detail in Ref. 16.

A second experiment was carried out by Lederman and collaborators in early 1957, essentially employing the β-decay of the μ^- particle, which was spin aligned during the decay of a π^- meson:

$$\pi^- \rightarrow \mu^- + \nu_\mu, \ \mu^- \rightarrow e^- + \bar{\nu}_e + \nu_\mu, \qquad (10.7)$$

and observing that many more electrons were emitted in one direction than the opposite direction. This experiment also confirmed that parity symmetry was violated in CC weak nuclear interactions.

These parity violating experiments indicated that the theta and tau mesons were indeed the same meson (later termed K^+) with different decay modes. In 1947 a new neutral particle of similar mass as the K^+ meson was discovered by Rochester and Butler using a cloud chamber exposed to cosmic rays. This particle was initially called a V-particle because upon decay it displayed two tracks corresponding to charged particles forming a V. The particle was a neutral K-meson (K^0) decaying to a π^+ and a π^- meson. In 1950, an event was found by Victor Hopper (1913-2005) and Sukumar Biswas (1924-2009) in which the positively charged track appeared to be a proton: this was the first evidence of a Λ^0 hyperon decaying into a proton and a π^- meson.

As discussed in Section 4.2, the Brookhaven Cosmotron was producing a significant number of V-particles by 1952. It was found that these particles were produced only in pairs, a typical strong nuclear reaction being:

$$\pi^- + p^+ \rightarrow \Lambda^0 + K^0 . \tag{10.8}$$

On the other hand, the V-particles decayed individually in about 10^{-12}s, which is about 10^{12} times longer than expected if the production process and decay mechanism are governed by the same kind of interaction.

This paradox of the strange V-particles was resolved by the introduction by Gell-Mann, Nishijima and collaborators, of the strangeness (S) additive quantum number, which was assumed to be conserved in strong nuclear interactions but not necessarily so in CC weak nuclear decay processes. The Λ^0 hyperon was allotted $S = -1$, while the K^0 meson was allotted $S = +1$. The non-strange particles, such as the charged pions, were allotted $S = 0$. Thus the decay

$$K^0 \rightarrow \pi^+ + \pi^- \tag{10.9}$$

does not conserve strangeness and proceeds only via a CC weak nuclear interaction process.

In 1955 Gell-Mann and Pais considered the behavior of neutral particles under the charge-conjugation operator C. In particular, they considered the K^0 meson and realized that unlike the photon and the neutral pion, which transform into themselves under the C operator so that they are their own antiparticles, the antiparticle of the K^0 $(S = +1)$, $\bar{K}^0$, was a distinct particle, since it had a different strangeness quantum number $(S = -1)$. They concluded that the two neutral mesons, K^0 and $\bar{K}^0$, are degenerate particles that exhibit unusual properties, since they can transform into each other via CC weak nuclear interactions such as

$$K^0 \rightleftharpoons \pi^+\pi^- \rightleftharpoons \bar{K}^0 . \tag{10.10}$$

In order to treat this novel situation, Gell-Mann and Pais suggested that it was more convenient to employ different particle states, rather than K^0 and $\bar{K}^0$, to describe neutral kaon decay. They suggested the following representative states:

$$K_1^0 = (K^0 + \bar{K}^0)/\sqrt{2}, \quad K_2^0 = (K^0 - \bar{K}^0)/\sqrt{2}, \qquad (10.11)$$

and concluded that these particle states must have different decay modes and lifetimes. In particular they concluded that K_1^0 could decay to two charged pions, while K_2^0 would have a longer lifetime and more complex decay modes. This conclusion was based upon the conservation of C in CC weak nuclear interaction processes: both K_1^0 and the $\pi^+\pi^-$ system are even (i.e. C = +1) under the C operation.

The particle-mixing theory of Gell-Mann and Pais was confirmed in 1957 by experiment, in spite of the incorrect assumption of C invariance in CC weak nuclear interaction processes (C violation was only discovered in 1957): the particle-mixing theory predicts that one-half of all K^0 decays should occur via K_2^0 and this was confirmed in 1957 by experiment. The particle-mixing theory led to a suggestion by Lev Landau (1908-1968) that the CC weak nuclear interactions may be *invariant* under the combined operation CP, although C and P are individually maximally violated.

Landau's suggestion implied that the Gell-Mann and Pais model of neutral kaons would still apply if the states, K_1^0 and K_2^0, were eigenstates of CP with eigenvalues +1 and −1, respectively. Since the charged pions had intrinsic parity $P_\pi = -1$, it was clear that only the K_1^0 state could decay to two charged pions, if CP was conserved, while the K_2^0 state would have a longer lifetime and more complex decay modes. It should be noted that this argument assumes that the intrinsic parities of the K^0 and $\bar{K}^0$ mesons are both even.

The suggestion of Landau was accepted for several years, since it nicely restored some degree of symmetry in CC weak nuclear interaction processes. However, as mentioned in Section 9.2, the 1964 surprising discovery by Cronin, Fitch and collaborators of the decay of the long-lived neutral K_2^0 meson to two charged pions led to the conclusion that CP is violated in this CC weak nuclear interaction.

The observed violation of CP conservation turned out to be very small ($\approx 0.2\%$) compared with the maximal violations ($\approx 100\%$) of both P and C conservation separately. Indeed the very smallness of the apparent CP violation led to a variety of suggestions explaining it in a CP-conserving way. However, these efforts were unsuccessful and CP violation in CC weak nuclear interactions was accepted within the framework of the SM.

An immediate consequence of this observation was that the role of K_1^0 (CP = +1) and K_2^0 (CP = −1), defined by Eq. (10.11) was replaced by two new particle states, corresponding to the short-lived (S) and long-lived (L) neutral kaons:

$$K_S^0 = (K_1^0 + \epsilon K_2^0)/(1 + |\epsilon|^2)^{\frac{1}{2}}, \quad K_L^0 = (K_2^0 + \epsilon K_1^0)/(1 + |\epsilon|^2)^{\frac{1}{2}}, \quad (10.12)$$

where the small complex parameter ϵ is a measure of CP impurity in the eigenstates K_S^0 and K_L^0. This method of describing CP violation in the SM, by introducing mixing of CP eigenstates was called 'indirect CP violation'. It is essentially a phenomenological approach with the parameter ϵ to be determined by experiment and does not provide any physical explanation for the origin of the CP violation.

It will now be shown to be plausible that the two pion decay of the long-lived neutral kaon discovered by Cronin, Fitch and collaborators in 1964 may be described, within the framework of the GM, in terms of physical quark mixing, assuming the *conservation of CP*. For simplicity I shall restrict the discussion to only the first two generations of quarks, since an extension to include all three generations would only have a small numerical impact upon the results.

I now turn to the framework of the GM in which the parity of the charged pions is *different* from $P_\pi = -1$, as in the SM. In the GM, the charged pions are composed of *weak eigenstate quarks* (see Sections 6.2 and 9.2) rather than mass eigenstate quarks and their corresponding antiparticles. Thus, while in the SM the π^- meson consists of a down quark (d) and an up antiquark ($\bar{u}$), i.e. $\pi^- \equiv [d\bar{u}]$, in the GM the π^- meson consists of a down-like weak eigenstate quark (d') and an up-like weak eigenstate antiquark ($\bar{u}$), i.e, $\pi^- \equiv [d'\bar{u}]$. It should be noted that by convention, the mass eigenstate up quark is equivalent to the weak eigenstate up quark in both the GM and the SM. In the two generation approximation, the weak eigenstate quark (d') is given approximately by

$$d' = d\cos\theta_c + s\sin\theta_c, \quad (10.13)$$

where $\cos\theta_c \approx 0.9742$ [see Eq. (10.1)]. Thus the π^- meson has approximately the structure:

$$\pi^- \equiv [d'u] = [d\bar{u}]\cos\theta_c + [s\bar{u}]\sin\theta_c. \quad (10.14)$$

The parity of the π^- meson can be denoted by

$$P_\pi = P_{d'}P_{\bar{u}} \equiv P_{d'}, \quad (10.15)$$

where

$$P[d'] = P_{d'}[d'] = P_d[d] \cos \theta_c + P_s[s] \sin \theta_c, \qquad (10.16)$$

so that the π^- meson exists in a mixed parity state: $\cos^2 \theta_c P_d + \sin^2 \theta_c P_s$, i.e. $\approx 95\% P_d + 5\% P_s$, with $P_d = -1$ and $P_s = +1$ (see Section 10.1). It should be noted that $P_{d'}$ is *not* an eigenvalue of the parity operator but is simply a short-hand notation for the linear superposition of mixed-parity eigenvalues, P_d and P_s, in Eq. (10.15). The mixed-parity parameter $P_{d'}$ may be treated analogously to an eigenvalue of a parity eigenstate, although unlike an eigenvalue, $[P_{d'}]^2$ is $\neq 1$.

Similarly, both the neutron $[ud'd']$ and the proton $[uud']$ exist in mixed-parity states: one has $P_n = P_u P_{d'} P_{d'} = -P_{d'} P_{d'}$ and $P_p = P_u P_u P_{d'} = P_{d'}$ for the neutron and proton parities, respectively.

I shall now demonstrate that the 1954 experiment of Chinowsky and Steinberger is compatible with the mixed-parity nature of the π^- meson, essentially determined by $P_{d'}$.

For the reaction $\pi^- + D \to 2n$, employed by Chinowsky and Steinberger, one requires the conservation of both parity and total angular momentum.

As indicated earlier, the spins of the π^- meson and the deuteron are known to be 0 and 1, respectively, and it is also known that the pion is captured by the deuteron from an S state so that the relative orbital angular momentum in the initial state is $l_i = 0$, and consequently the total angular momentum of the initial state is $j_i = 1$. Thus the overall parity of the initial state is given by

$$(-1)^{l_i} P_\pi P_D = (-1)^0 P_{d'} P_n P_p = P_{d'}(-P_{d'} P_{d'}) P_{d'} = -[P_{d'}]^4. \qquad (10.17)$$

The overall parity of the final state for relative orbital angular momentum l_f is

$$(-1)^{l_f} P_n P_n = (-1)^{l_f}(-P_{d'} P_{d'})(-P_{d'} P_{d'}) = (-1)^{l_f}[P_{d'}]^4. \qquad (10.18)$$

Again, as discussed earlier, since the neutrons are identical fermions, the only allowed antisymmetric state of the two neutrons with $j_f = 1$ (conservation of total angular momentum) is a $^3P_1$ state, so that $l_f = 1$, and hence the parity of the final state is the same as the parity of the initial state, $-[P_{d'}]^4$. Thus, although parity is conserved in the reaction, the mixed-parity of the π^- meson, $P_{d'}$, is not determined by the experiment.

Another important consequence of the mixed-parity of the charged pions is that it provides a quantative description of the decay of the long-lived K_2^0 meson into two charged pions: $K_2^0 \to \pi^+ + \pi^-$, as discovered by Cronin,

Fitch and collaborators in 1964, *without the violation of CP symmetry* in the CC weak nuclear interaction process [38].

As discussed earlier within the framework of the SM, the decay $K_2^0 \to \pi^+ + \pi^-$ indicated violation of the CP symmetry in the CC weak nuclear interaction process, since the K_2^0 meson is considered to exist in a CP $= -1$ eigenstate, while the final two charged pion state is a CP $= +1$ eigenstate. However, this conclusion depends critically upon the charged pion parity being $P_\pi = -1$.

In the GM, mesons are composed of weak eigenstate quarks, which in the two generation approximation indicates that the two weak eigenstate down-like quarks, d' and s' are given by

$$d' = d \cos \theta_c + s \sin \theta_c, \quad s' = s \cos \theta_c - d \sin \theta_c, \tag{10.19}$$

respectively.

Inserting the physical quark states, d' and s', into the structures of $K^0 \equiv [d'\bar{s}']$ and $\bar{K}^0 \equiv [\bar{d}'s']$ using Eq. (10.19), one can expand the state K_2^0:

$$\begin{aligned}
K_2^0 &= \frac{1}{\sqrt{2}} (K^0 - \bar{K}^0) \\
&= \frac{1}{\sqrt{2}} ([d'\bar{s}'] - [\bar{d}'s']) \\
&= \frac{1}{\sqrt{2}} ([(d \cos \theta_c + s \sin \theta_c)(\bar{s} \cos \theta_c - \bar{d} \sin \theta_c)] \\
&\quad - [(\bar{d} \cos \theta_c + \bar{s} \sin \theta_c)(s \cos \theta_c - d \sin \theta_c)]) \\
&= \frac{1}{\sqrt{2}} ([d\bar{s}] - [\bar{d}s]). \tag{10.20}
\end{aligned}$$

Thus in the GM the cancellation of quark amplitudes leaves exactly the same pure quarks as those in the SM, so that the state K_2^0 remains a CP $= -1$ eigenstate.

Inserting the physical quark state d' into the structure of the charged 2π system:

$$\begin{aligned}
\pi^+ \pi^- &= [u\bar{d}'][d'\bar{u}] \\
&= [(u)(\bar{d} \cos \theta_c + \bar{s} \sin \theta_c)][(d \cos \theta_c + s \sin \theta_c)(\bar{u})] \\
&= [u\bar{d}][d\bar{u}] \cos^2 \theta_c + [u\bar{s}][s\bar{u}] \sin^2 \theta_c + [u\bar{s}][d\bar{u}] \sin \theta_c \cos \theta_c \\
&\quad + [u\bar{d}][s\bar{u}] \sin \theta_c \cos \theta_c. \tag{10.21}
\end{aligned}$$

The u and d quarks have negative parity, while the s quark has positive parity, and their antiparticles have the corresponding opposite parities.

Thus the first two components of Eq. (10.21) are eigenstates of CP = +1, while the remaining two components, $[u\bar{s}][d\bar{u}]$ and $[u\bar{d}][s\bar{u}]$, with amplitude $\sin\theta_c \cos\theta_c$ are not individually eigenstates of CP. However, taken together, the state $([u\bar{s}][d\bar{u}] + [u\bar{d}][s\bar{u}])$ is an eigenstate of CP with eigenvalue CP = -1. This treatment of the two components together in order to create a CP eigenstate is equivalent to the same treatment of K^0 and $\bar{K}^0$ in the SM.

Taking the square of the product of the amplitudes of the two components comprising the CP = -1 eigenstate to be the 'joint probability' of those two states existing together simultaneously, one obtains that this probability is $(\sin\theta_c \cos\theta_c)^4 = 2.34 \times 10^{-3}$, using $\cos\theta_c = 0.9742$.

The existence of a small component of the $\pi^+\pi^-$ system with eigenvalue CP = -1 indicates that the long-lived K_2^0 meson can decay to the charged 2π system *without violating* CP conservation. Moreover, the estimated decay rate is in good agreement with experiment.

The conclusion from this observation is that CP is *conserved* in CC weak nuclear interactions, in agreement with Landau's earlier suggestion.

10.3   Parity of Neutral Pion

It is of interest to discuss the parity of the neutral pion within the framework of the GM, since the quark model of the π^0-meson involves mixed-quark d' states.

Direct determination of the parity of the π^0-meson presents a difficult experimental challenge. In 1950 Yang indicated that the parity of the neutral pion could in principle be determined by measuring the polarizations of the two photons in the decay $\pi^0 \rightarrow \gamma\gamma$, although no method has yet been developed to measure the polarization of the high energy (≈ 70 MeV) photons involved in the decay. Yang showed that the polarizations of the two photons are perpendicular or parallel depending whether the π^0 is *pseudoscalar* or *scalar*, respectively. However, it was soon pointed out that the double Dalitz decay $\pi^0 \rightarrow e^+e^-e^+e^-$, which proceeds through an intermediate state of two virtual photons, is sensitive to the parity of the neutral pion, since the plane of a Dalitz pair is correlated with the polarization of the associated virtual photon.

In the SM, the parity of the neutral pion is assumed to be $P_0 = -1$. In 1954 Chinowsky and Steinberger established, within the theoretical framework of the time, that the negative pion was pseudoscalar, i.e. had $P_\pi = -1$. In 1955 they determined that the branching ratio between the capture reactions in deuterium: $\pi^- + D \to 2n + \pi^0$ and $\pi^- + D \to 2n + \gamma$, was less than 0.1% and showed that this provided strong evidence that the neutral pion was also pseudoscalar.

Early attempts during 1959-1962 to determine directly the parity of the neutral pion via its double Dalitz decay were only sufficiently accurate to confirm that the neutral pion was a pseudoscalar ($P_0 = -1$) rather than a scalar ($P_0' = +1$) particle. An underlying assumption in the analyses of these experiments was that the π^0-meson had a definite intrinsic parity $P_0 = +1$ or $P_0 = -1$.

Recently, a new determination of the parity of the π^0-meson, employing the double Dalitz decay technique was carried out in 2008 [39]. Analysis of this experiment concluded that the neutral pion had $P_0 = -1$, and placed a limit on scalar contributions to the $\pi^0 \to e^+e^-e^+e^-$ decay amplitude of less than 3.3%, assuming CPT conservation.

In the SM the π^0-meson, according to the quark model, has the substructure

$$\pi^0 = ([u\bar{u}] - [d\bar{d}])/\sqrt{2}, \tag{10.22}$$

so that, assuming that the elementary quarks have $P_q = +1$ while their corresponding antiquarks have parity $P_{\bar{q}} = -1$, the neutral pion has parity $P_0 = -1$. Consequently, the SM predicts that the scalar contribution to the total decay amplitude is zero.

In the GM, involving mixed-quark states d' rather than pure quark states d, the π^0-meson has, in the two generation approximation, the structure:

$$\pi^0 = ([u\bar{u}] - [d'\bar{d}'])/\sqrt{2}, \tag{10.23}$$

where

$$d' = d\cos\theta_c + s\sin\theta_c, \quad \bar{d}' = \bar{d}\cos\theta_c + \bar{s}\sin\theta_c, \tag{10.24}$$

and

$$\pi^0 = ([u\bar{u}] - [d\bar{d}]\cos^2\theta_c - [s\bar{s}]\sin^2\theta_c \tag{10.25}$$
$$- ([d\bar{s}] + [s\bar{d}])\sin\theta_c\cos\theta_c)/\sqrt{2}.$$

In the GM, the u-quark and the d-quark both consist of two rishons and one antirishon, so that they have parity $P_u = -1$ and $P_d = -1$, respectively, while the s-quark consists of three rishons and two antirishons and has parity $P_s = +1$. The antiparticles of these three quarks have the corresponding opposite parities: $P_{\bar{u}} = +1$, $P_{\bar{d}} = +1$ and $P_{\bar{s}} = -1$.

Furthermore, defining the phase of the charge-conjugation operator (C) by the relation: $C|q\rangle = +|\bar{q}\rangle$, it is seen that the first three components of Eq. (10.25) are eigenstates of CP with eigenvalue CP = -1, while the remaining two components, $[d\bar{s}]$ and $[s\bar{d}]$, with amplitude $\sin\theta_c \cos\theta_c/\sqrt{2}$ are not individually eigenstates of CP. However, taken together, the state $([d\bar{s}] + [s\bar{d}])$ is an eigenstate of CP with eigenvalue CP = $+1$.

It should be noted that each of the components with eigenvalue CP = -1 also has eigenvalues C = $+1$ and P = -1 eigenstates, while the state with eigenvalue CP = $+1$ also has eigenvalues C = $+1$ and P = $+1$ eigenstates, so that the pseudoscalar or scalar nature of the neutral pion is determined by its existence in either CP = -1 or CP = $+1$ eigenstates, respectively.

Taking the square of the product of the amplitudes comprising the CP = $+1$ eigenstates to be the 'joint probability' of these two states existing together simultaneously, one can calculate that this probability is $\frac{1}{4}\sin^4\theta_c \cos^4\theta_c = 5.84 \times 10^{-4}$, using $\cos\theta_c = 0.9742$.

Analysis of the data obtained in the 2008 experiment carried out [39] at the Fermi Laboratory, found that the scalar contributions to the decay *amplitude* were less than 3.3%. In the GM, since CP is conserved in the electromagnetic decay, the occurrence of scalar contributions to the decay amplitude arises entirely from the joint probability *amplitude* of the CP = $+1$ eigenstate component of the neutral pion. This is given by $\frac{1}{2}\sin^2\theta_c \cos^2\theta_c = 2.42 \times 10^{-2}$. Normalizing to the total decay amplitude, one finds that the scalar contribution to the decay amplitude is $\approx 2.5\%$, which is compatible with the experimental result of less than 3.3%. This experimental result is of the same order of magnitude as the GM prediction, suggesting that further experimentation may determine a non-zero scalar contribution to the decay amplitude.

Chapter 11

Electroweak Connection

11.1 Introduction

One of the cornerstones of the SM [2–4, 6, 7] is the treatment of the electromagnetic interaction and the two classes of weak interactions: (i) the CC weak nuclear interaction involving the W^+ and W^- bosons and (ii) the neutral weak nuclear interaction involving the Z^0 boson, in terms of a $U(1) \times SU(2)_L$ local gauge theory, known as the Glashow, Weinberg and Salam (GWS) model.

The GWS model leads to a relation between the electromagnetic and the weak nuclear interactions, which has been termed the 'electroweak connection': *the charge-preserving weak interaction is completely fixed by the electromagnetic interaction and the charge-changing weak interaction* [24]. In this sense, the electromagnetic interaction involving the massless neutral photons and the CC weak nuclear interaction involving the massive charged W bosons and the neutral weak nuclear interaction involving the massive neutral Z boson are *related* but are *not strictly unified*, since the relationship involves two independent coupling constants, an electric charge Q and a weak 'charge' g_w. This GWS model has been discussed earlier within the framework of the SM in some detail (see Section 5.2).

The above electroweak connection, which was derived during the 1960s by Glashow, Weinberg and Salam, is in excellent agreement with the experimental data. However, in 2008 during the development of the GM, I noted that the derivation of the electroweak connection, within the framework of the SM, suffered from a number of problems.

First, the SM requires the existence of a scalar Higgs field to spontaneously break the assumed $U(1) \times SU(2)_L$ local gauge symmetry. This implied the existence of a new massive spin zero boson, the Higgs boson, which in 2008 had not yet been detected.

Second, the SM requires the Higgs field to couple to the originally massless fermions, the leptons and quarks, to produce their finite masses, in a manner which does not violate the assumed gauge invariance. In the SM this requires the fermion-Higgs coupling strength to be dependent upon the mass of the fermion, so that a new parameter is required for each fermion mass in the theory.

Third, as pointed out by Wilczek in 2005, the requirement of a Higgs field, which fills the whole of space, leads to a cosmological term in the general theory of relativity that is much larger than is allowed by observations [20].

Consequently, following a brief summary of the development of the electroweak connection within the framework of the SM, I shall indicate how all the above problems may be avoided by assuming that the weak nuclear interaction is *not* a fundamental interaction, arising from a local gauge invariance within the framework of the GM.

11.2 The SM and the Electroweak Connection

In the SM derivation of the electroweak connection, the symmetry involved, namely $U(1) \times SU(2)_L$, is more complex than the original $U(1)$ symmetry developed by Weyl, Fock and London for the electromagnetic interaction, corresponding to a change in the phase of the wave function describing the electron field, so that the local gauge principle associated with QED was required to be extended to the non-Abelian case of $SU(2)$, as studied by Yang and Robert Mills (1927-1999).

In 1954 Yang and Mills proposed that the strong nuclear interaction may be described by a field theory similar to electromagnetism that is exactly gauge invariant. They attempted to do this by postulating that the local gauge group was the $SU(2)$ strong isospin group, introduced by Heisenberg in 1932 (see Section 4.2).

Unfortunately, this idea of Yang and Mills failed, primarily because the mediating particles associated with the strong nuclear forces required to bind the nucleons together within a nucleus were *massive* unlike the photon of the QED case: the Yang-Mills gauge field required that the mediating particles must be *massless*.

It should be noted that in 1973, the chromodynamic strong force was developed into QCD, which is associated with an $SU(3)$ symmetry and is mediated by massless gluons within the framework of the SM. More recently, the chromodynamic force was proposed to be the strong force, which binds together the rishon and antirishon constituents of the leptons and quarks, within the framework of the GM: again the mediating particles are massless hypergluons (see Chapter 6).

The SM derivation of the electroweak connection is based upon two main ideas: (i) nature exhibits an $SU(2)$ symmetry associated with the CC weak nuclear interaction, as indicated by experimental observations of CC weak nuclear interaction processes; and (ii) the CC weak nuclear interaction is a *fundamental* interaction on a par with the electromagnetic interaction. This led to the notion that the CC weak nuclear interaction is a consequence of an $SU(2)$ *local* gauge transformation, analogous to the electromagnetic interaction, which obeys a $U(1)$ local gauge transformation.

Furthermore, the measurement of the helicity of the electron neutrino in 1958 by Goldhaber and collaborators to be *negative*, so that the electron neutrino is left-handed, indicated that the CC weak nuclear interaction only involves left-handed fermions and right-handed antifermions, if the electron neutrino is not massless as was assumed by several scientists at that time. In this case, the local gauge would be termed $SU(2)_L$, i.e. left-handed $SU(2)$ symmetry.

The electroweak connection has been discussed in some detail in Section 5.2. During the 1950s and the 1960s, physicists were interested in trying to understand the relationship or possible unification of the electromagnetic interaction and the CC weak nuclear interaction. During a ten year period, several attempts were made to find such a relationship within the framework of the SM.

First, in 1957 Schwinger suggested that the photon (γ) and the W^+ and W^- bosons form a triplet in an $SU(2)$ weak isospin *local* gauge theory. This suggestion suffered from the fact that the large masses of the W bosons, required to account for the very short-range nature of the CC weak nuclear interaction, had to be inserted 'by hand' into the theory, in conflict with the local gauge invariance requirement that the mediating particles should be *massless*.

Second, in 1958 Bludman proposed that many aspects of the CC weak nuclear interaction could be described in terms of a triplet of three vector bosons, W^+, W^0 and W^-, in a weak isospin space, by an $SU(2)$ gauge theory, provided that it was of the *global* type, thereby avoiding the mass problem. This implied that the CC weak nuclear interaction was *not* a fundamental interaction, as was generally assumed in the SM.

Third, as discussed in Section 5.2, in 1961 Glashow proposed that the CC weak nuclear interaction, considered as a fundamental force, could be associated with the fundamental electromagnetic force, if Bludman's triplet of vector bosons, (W^+, W^0, W^-) was associated with a singlet vector boson, B^0, and the two neutral bosons 'mixed' in such a way that they produced a massive Z^0 boson and the massless photon (γ):

$$\gamma = B^0 \cos\theta_W + W^0 \sin\theta_W, \qquad (11.1)$$

$$Z^0 = -B^0 \sin\theta_W + W^0 \cos\theta_W, \qquad (11.2)$$

where θ_W is the electroweak mixing angle. Such a mixing was possible in quantum mechanics because the two neutral bosons, W^0 and B^0, possessed the same quantum numbers [3]. This was a major step in the development of the electroweak connection and led to the discovery of the Z^0 particle in 1983.

Unfortunately, the above proposal of Glashow also suffered from the fact that the large masses of the W^+, Z^0 and W^- bosons had to be inserted into the theory by hand. However, this mass problem was resolved independently by Weinberg and Salam, in 1967 and 1968, respectively, by employing the notion of *spontaneous symmetry breaking* (see Section 5.2), involving the Higgs mechanism developed by Higgs, Brout and Englert, so that the W and Z bosons acquired mass, while the photon remained massless. This GWS model does provide a derivation of the electroweak connection within the framework of the SM [24, 25].

On the other hand, it should be noted that Lyre has considered the concept of spontaneous symmetry breaking in the electroweak gauge theory of the SM via the so-called Higgs mechanism [33]. Lyre concludes that transforming the Higgs Lagrangian in a gauge-invariant manner results in a mere reshuffling of the degrees of freedom, and is comparable to that of a coordinate transformation. Consequently, it is unrealistic to consider that the Higgs mechanism can dynamically generate the masses of the elementary particles of the SM.

11.3   The GM and the Electroweak Connection

I shall now indicate how all the problems associated with the derivation of the electroweak connection in the SM may be avoided, if the weak nuclear interaction, mediated by the W and Z bosons, is assumed *not* to be a fundamental interaction, as in the SM.

The GM has two important differences from the SM. First it assumes that the weak nuclear interaction, mediated by the W and Z bosons, is *not* a fundamental interaction so that any SU(2) symmetry associated with it will be a *global* gauge symmetry, rather than a local gauge symmetry as in the SM. Second the GM assumes that the leptons, quarks and the W and Z bosons are all composite particles, consisting of massless rishons and/or antirishons, so that each of these particles is left-handed or right-handed, depending upon the number of left-handed antirishons within each fermion or boson.

In particular, both the electron and the electron neutrino are left-handed, since they are composed of three antirishons. These properties of the fermions give rise to the left-handed nature of the $SU(2)$ symmetry, i.e. $SU(2)_L$, described earlier (see Section 5.2): indeed, the GM *predicts* the left-handed nature of the weak isospin $SU(2)$ symmetry. Earlier this had been a problem when neutrinos were found not to be massless, as previously assumed, and yet the electron neutrino was measured to have negative helicity and hence was left-handed.

As mentioned in Section 4.3, the first attempt at a theory of the CC weak nuclear force involved in the β-decay of a neutron into a proton was made by Fermi in 1934. Fermi described the β-decay process in terms of two interacting vector currents, analogous to the Dirac electromagnetic current:

$$j_{em}^\mu = \bar{\psi}\alpha_\mu\psi, \qquad (11.3)$$

where ψ is the electron field and α_μ are Dirac matrices described in Section 3.3, so that the matrix element describing the process could be written as:

$$M = \frac{F}{\sqrt{2}}j_1^\mu j_2^\mu, \qquad (11.4)$$

where F is the Fermi weak coupling constant and j_1^μ and j_2^μ are given by

$$j_1^\mu = \bar{\psi}_p\alpha_\mu\psi_n, \quad j_2^\mu = \bar{\psi}_\nu\alpha_\mu\psi_e. \qquad (11.5)$$

Here $\bar{\psi} = \psi^\dagger\alpha_0$, so that the matrix products involving $\psi^\dagger$ and ψ have Lorentz transformation invariant properties [22].

In the GM, the β-decay of a neutron into a proton, each of the two interacting currents j_1^μ and j_2^μ is associated with a well-defined weak isospin doublet set of fermion fields involving a well-defined particle number p (see Table 6.1).

In the GM, the CC weak nuclear interaction, mediated by W^+ and W^- bosons acts between the two particles comprising the known six weak isospin doublets in a 'universal' manner: (ν_e, e^-), (ν_μ, μ^-), (ν_τ, τ^-), (u, d), (c, s), (t, b). Here the first particle and the second particle within a doublet has $i_3 = +\frac{1}{2}$ and $i_3 = -\frac{1}{2}$, respectively, and the charge of each particle is given by

$$Q = i_3 + \frac{1}{2}p, \tag{11.6}$$

where $p = -1$ for leptons and $p = \frac{1}{3}$ for quarks (see Table 6.1).

Each weak isospin doublet may be written as a doublet of complete Dirac spinors, e.g. the left-handed doublet (ν_e, e^-):

$$\chi = \Gamma \begin{pmatrix} \nu_e \\ e^- \end{pmatrix}, \tag{11.7}$$

where Γ is the usual operator, which projects out left-handed particle states (see Eq. (5.22)). It should be noted that all the weak isospin doublets are left-handed particles, except for (ν_μ, μ^-) and (c, s), which are effectively right-handed antiparticles, in the GM.

It should be noted that in the following derivation of the electroweak connection, it is only necessary to consider one weak isospin doublet, since each pair contributes additively and independently to each total current and the CC weak nuclear interaction is the same for each case.

In 1961 Glashow attempted to 'unify' the CC weak nuclear and the electromagnetic interactions into a single gauge theory (see Section 11.2). Glashow proposed that this *electroweak interaction* was of the form:

$$H_{ew} = g_w \mathbf{j}^\mu . \mathbf{W}_\mu + g'_w j_0^\mu B_\mu, \tag{11.8}$$

where the first term with weak charge g_w corresponds to an $SU(2)_L$ symmetry and the second term with weak charge g'_w corresponds to an independent $U(1)$ symmetry. The current associated with the second term, j_0^μ, is required to be invariant in the weak isospin space possessing $SU(2)_L$ symmetry. The appropriate current for j_0^μ is $(j_{em}^\mu - j_3^\mu)$ [24], rather than the current originally assumed by Glashow, which required the need to employ different quantum numbers for left-handed and right-handed particles [25]. Thus the complete electroweak interaction is:

$$H_{ew} = g_w \mathbf{j}^\mu . \mathbf{W}_\mu + g'_w (j_{em}^\mu - j_3^\mu) B_\mu. \tag{11.9}$$

In the GM, it is expected that this electroweak interaction, deduced primarily from experiment, will be invariant under a $U(1)_p \times SU(2)_L$ *global gauge* transformation, corresponding to the conservation of both particle number p and weak charge g_w. Thus, the interaction H_{ew} should be invariant under the global gauge transformation:

$$\chi \rightarrow \chi' = \exp(i\lambda p) \exp(i\mathbf{\Lambda} . \tau \Gamma)\chi, \tag{11.10}$$

where λ and Λ are an arbitrary constant and an arbitrary constant vector, respectively.

From the interaction (11.9) it is quite straight forward to derive the electroweak connection. For simplicity the lepton pair (ν_e, e^-) will be considered. The total neutral (charge preserving) interaction is given by

$$H_{ew}^0 = g'_w B_\mu j_{em}^\mu + (g_w W_\mu^3 - g'_w B_\mu) j_3^\mu \,. \tag{11.11}$$

For the neutral electron neutrino, only the second term involving j_3^μ contribute, so that this must be associated with the weak boson Z^0, i.e.

$$Z_\mu = (g_w W_\mu^3 - g'_w B_\mu)/(g_w^2 + g_w'^2)^{\frac{1}{2}} \,, \tag{11.12}$$

where a normalization factor $(g_w^2 + g_w'^2)^{-\frac{1}{2}}$ has been inserted so that the states created and destroyed by Z_μ (and A_μ) are normalized in the same manner as those of B_μ and W_μ^3. From Eqs. (11.1) and (11.2), it is seen that the electroweak mixing angle is given by

$$\sin\theta_W = g'_w/(g_w^2 + g_w'^2)^{\frac{1}{2}} \,, \quad \cos\theta_W = g_w/(g_w^2 + g_w'^2)^{\frac{1}{2}} \,. \tag{11.13}$$

The orthogonal linearly independent combination, corresponding to Eq. (11.12), is the electromagnetic interaction:

$$A_\mu = B_\mu \cos\theta_W + W_\mu^3 \sin\theta_W \,. \tag{11.14}$$

From Eq. (11.14) one has, equating the associated currents and coupling constants:

$$e j_{em}^\mu = g'_w (j_{em}^\mu - j_3^\mu) \cos\theta_W + g_w j_3^\mu \sin\theta_W \,, \tag{11.15}$$

so that

$$e = g'_w \cos\theta_W = g_w \sin\theta_W \,. \tag{11.16}$$

Furthermore, if one writes the weak neutral interaction as $g_Z j_{NC}^\mu Z_\mu$, then from Eq. (11.12):

$$g_Z j_{NC}^\mu = g_w j_3^\mu \cos\theta_W - g'_w (j_{em}^\mu - j_3^\mu) \sin\theta_W \,, \tag{11.17}$$

so that

$$g_Z j_{NC}^\mu = g_w (j_3^\mu - j_{em}^\mu \sin^2\theta_W)/\cos\theta_W \,. \tag{11.18}$$

Thus the complete electroweak interaction has the form:

$$\begin{aligned} H_{ew} = {}& e j_{em}^\mu A_\mu + g_w (j_1^\mu W_\mu^1 + j_2^\mu W_\mu^2) \\ & + g_w (j_3^\mu - j_{em}^\mu \sin^2\theta_W) Z_\mu/\cos\theta_W \,, \end{aligned} \tag{11.19}$$

which gives the electroweak connection: the neutral weak interaction mediated by the Z^0 bosons is completely determined by the electromagnetic and charge-changing interactions, and their coupling constants e and g_w. In addition,the relative masses of the W and Z bosons are given in terms of the electroweak mixing angle:

$$M_W = M_Z \cos\theta_W .\qquad(11.20)$$

In the GM, the CC weak nuclear interaction is treated as an effective interaction, arising from residual interactions of the strong nuclear interaction, which binds the constituents of leptons and quarks together, so that the CC weak nuclear interaction is *not* a fundamental interaction arising from a local gauge invariance. It has been found that a *global* $U(1) \times SU(2)_L$ gauge invariance, corresponding to the conservation of particle number p and weak charge g_w, is sufficient to determine the observed electroweak connection.

11.4   Weak Bosons

According to the electroweak connection of the GWS model, the electroweak interactions are mediated by four bosons: W^+, W^-, Z^0 and γ. This result was derived in Section 11.2 within the framework of the SM, assuming that the weak bosons W and Z mediated fundamental weak nuclear interactions and satisfied a weak isospin local left-handed gauge symmetry, $SU(2)_L$, so that the electroweak interactions involved a $U(1) \times SU(2)_L$ local gauge symmetry.

Unfortunately, the original version of the electroweak connection, proposed in 1961 by Glashow, suffered from the fact that the large masses of the W and Z bosons had to be inserted into the theory 'by hand', since the local gauge theory required that the mediating particles of the electroweak interactions should be massless for consistency.

However, as indicated in Section 5.2, this mass problem was resolved independently by Weinberg and Salam in 1967 and 1968, respectively, by employing the notion of spontaneous symmetry breaking involving the so-called Higgs mechanism, developed in 1964 by Higgs, Brout and Englert, so that the W and Z bosons acquired mass, while the photon remained massless.

Unfortunately, Lyre [33] in 2008 concluded that transforming the Higgs Lagrangian in a gauge-invariant manner results in a mere reshuffling of the degrees of freedom, comparable to a coordinate transformation, so that it is unrealistic to consider that the Higgs mechanism can dynamically generate the masses of the elementary particles of the SM.

The GM has two important differences from the SM. First, it assumes that the weak nuclear interaction, mediated by the W and Z bosons, is *not* a fundamental interaction so that any $SU(2)$ symmetry associated with it will be a *global* gauge symmetry, rather than a local gauge symmetry as in the SM: this implies that there is no mass problem as found by Glashow. Second, the GM assumes that the leptons, quarks and the W and Z bosons are all composite particles, consisting of massless rishons and/or antirishons, so that each of these particles is left-handed or right-handed, depending upon the number of left-handed antirishons within each fermion or boson: these properties of the fermions give rise to the left-handed nature of the $SU(2)$ symmetry, i.e $SU(2)_L$.

In the GM, the W^+ and W^- weak bosons are composite colorless particles with the rishon structures $TTT\bar{V}\bar{V}\bar{V}$ and $\bar{T}\bar{T}\bar{T}VVV$, respectively, and each has spin and parity 1^-. Since these massive W bosons mediate the CC weak nuclear interaction acting between the six known weak isospin doublets, (ν_e, e^-), (ν_μ, μ^-), (ν_τ, τ^-), (u, d), (c, s), (t, b), whose members have the *same* intrinsic parity, these interactions are expected to be approximately 100% parity violating, as observed.

In the GM, the rishon structure of the Z^0 boson is rather more difficult to discern. However, there are several properties of this particle that are known within the framework of the GM. First, it has to be a *colorless* particle. Second, it should be closely associated with the W bosons, so that it probably also consists of a colorless set of three rishons and a colorless set of three antirishons. Third, its structure should *not* include a colorless rishon-antirishon pair that could immediately annihilate, e.g. $T_r\bar{T}_{\bar{r}}$.

After some consideration, I found that the most likely lowest mass rishon structure of the Z^0 boson was the following:

$$Z^0 \equiv TV\bar{T}\bar{V}\Pi, \quad \Pi = [\bar{V}^*V + \bar{V}V^*]/\sqrt{2}, \tag{11.21}$$

as in Eq. (6.13). Such a boson is expected to have spin and parity 1^-, provided the excited V^*-rishon or its excited $\bar{V}^*$-antirishon exists in a 2s state, as in the higher generations of the leptons and quarks.

The mass of the W boson is 80.4 GeV and that of the Z^0 boson is 91.2 GeV, obtained from measurements of the electroweak mixing angle θ_W from Eq. (11.20). In the GM, the mass of a particle with the annihilation rishon structure $TVV\bar{T}\bar{V}\bar{V}$ is expected to be less than that of the W boson, because of attraction between the $T\bar{T}$ rishon-antirishon pair. However, the existence of an excited V^*-rishon in a 2s state possibly allows the mass of the Z^0 boson to be larger than the W boson.

The above model of the Z^0 boson suggests the possible existence of excited Z^0 weak bosons. The next excited Z^0-type particle is expected to have the excited V^*-rishon and its excited $\bar{V}^*$-antirishon in a 2p state. In this case, the spin and parity of such a particle is expected to be 0^+, 1^+ or 2^+. It is interesting to note that the boson discovered at CERN in 2012 is claimed to have a spin and parity of 0^+. This particle has a mass of about 125 GeV, which could be in agreement with a Z^0-type particle with a V^*-rishon in an excited 2p state.

Chapter 12

Epilogue

In this book I have reported the research I undertook in 2001 to determine possible dubious assumptions that were made during the development of the Standard Model (SM) of particle physics. Such dubious assumptions may be responsible for the SM being considered to be incomplete, in the sense that it provides little understanding of several empirical observations such as the existence of three families of the elementary leptons and quarks that have similar properties apart from mass, while enjoying considerable success in describing the interactions of leptons and the multitude of hadrons with each other, as well as the decay modes of the unstable leptons and hadrons.

This research is described in the first four chapters, which provide a historical introduction covering three eras of progress in the understanding of the nature of the Universe in terms of its building blocks, i.e. the elementary particles of the constituent ordinary matter, and the nature of the forces acting between these elementary particles.

The three eras cover a period of over 2500 years: (i) the era of classical physics, which is assumed to run from antiquity (ca. 600 BC) until about 1895 and is associated with the macroscopic world in which only the gravitational and electromagnetic forces are evident from direct experience because of their long-range nature, and ordinary matter was considered to be composed of atoms; (ii) the era of transitional physics, 1895-1932, in which several discoveries were made, which could not be reconciled with classical physics and indicated the need for new physics; and (iii) the era of modern physics, 1932 to the present day, which is associated mainly with the microscopic (subatomic) world in which both the weak nuclear and the strong nuclear short-range forces operate.

The SM was essentially finalized in the mid-1970s following the experimental confirmation of quarks, so that the elementary particles of the

SM were considered to be six leptons and six quarks, following significant progress in the era of transitional physics by the development of two new theories, relativity theory and quantum theory, and also the development of new apparatus, including the Crookes tube and the Wilson cloud chamber. Further significant progress was achieved during the modern era following the development of powerful accelerators, leading to the creation of many new particles.

In the modern era, the elementary particles, the leptons and quarks, of the SM were classified in terms of additive quantum numbers based upon the notions of weak isospin and strong isospin, respectively. The six leptons were classified in terms of four additive quantum numbers, charge Q and three flavor lepton numbers, L_e, L_μ and L_τ, while the six quarks were classified in terms of six additive quantum numbers, charge Q, baryon number A and four flavor quantum numbers, strangeness S, charm C, bottomness B and topness T. Unfortunately, apart from electric charge, the leptons and quarks were classified in terms of different additive quantum numbers, so that the overall classification was essentially nonunified.

The SM in 2000 consisted of twelve elementary spin-$\frac{1}{2}$ particles, six leptons and six quarks, which surprisingly fell naturally into three families or generations: (i) ν_e, e^-, u, d ; (ii) ν_μ, μ^-, c, s ; (iii) ν_τ, τ^-, t, b . Each generation consists of two leptons with charges $Q = 0$ and $Q = -1$ and two quarks with charges $Q = +\frac{2}{3}$ and $Q = -\frac{1}{3}$. The masses of the particles increased significantly with each generation.

The existence of these three generations remained a mystery for many years but did indicate that the SM contained a mini-Mendeleev type periodic table. This suggested, in spite of no direct experimental evidence, that the elementary leptons and quarks were actually composite particles. This was supported by considerable indirect evidence, especially the exact equality of the magnitudes of the electric charges of the electron $Q = -1$ and the proton $Q = +1$, leading to atoms being electrically neutral. The SM is described in detail in Chapter 5.

During 2001 I concluded that a basic problem with the SM was its nonunified classification of its elementary leptons and quarks, since this presented a major stumbling block for the development of a composite model of these particles.

In 2002 I investigated the possibility of a new simpler and unified classification scheme of the leptons and quarks of the SM. Fortunately, I established such a possibility, which was based upon the use of only three additive quantum numbers: charge Q, particle number p and generation quantum number g. Furthermore, these three quantum numbers could be chosen so that they were conserved in all interactions, unlike several of the SM quantum numbers, S, C, B and T, which may undergo a change of one unit in weak interactions.

Another dubious assumption of the SM, discovered in 2002, concerns the method it employs to accommodate the universality of the CC weak nuclear interaction, mediated by W bosons.

In the SM, the observed universality of the CC weak nuclear interaction in the lepton sector is described by assuming that the mass eigenstate leptons form weak isospin doublets. The leptons have weak isospin-$\frac{1}{2}$, whose third component is related to both charge and lepton number. This means that each neutrino interacts with its corresponding charged lepton with the full strength of the CC weak nuclear interaction and does not interact with the other charged leptons. This is guaranteed by the conservation of flavor lepton numbers, L_e, L_μ and L_τ.

On the other hand, the universality of the CC weak nuclear interaction in the quark sector is treated differently in the SM. In 1963, Cabibbo proposed that in hadronic processes the CC weak nuclear interaction was shared between $\Delta S = 0$ and $\Delta S = 1$ transition amplitudes in the ratio $\cos \theta_c : \sin \theta_c$, where $\theta_c \approx 13°$, and obtained universal agreement for the various CC weak nuclear hadronic interaction processes. However, this proposal of Cabibbo is very dubious, since it is based upon the non-conservation of the flavor strangeness quantum number S, which if conserved would guarantee that the strange quark interacts with the full strength of the CC weak nuclear interaction with its corresponding weak isospin doublet partner, the charmed quark.

In the GM, it is assumed that the six quarks form three weak isospin doublets: (u, d), (c, s) and (t, b) analogous to the six leptons, which form the three weak isospin doublets: (ν_e, e^-), (ν_μ, μ^-) and (ν_τ, τ^-). This overcomes the dubious assumption of Cabibbo concerning the universality of the CC weak nuclear interaction, which assumes that the weak isospin doublets of quarks are (u, d'), (c, s') and (t, b'), where d', s' and b' are CKM mixed-quark states. The universality of the CC weak nuclear interaction in the GM is guaranteed by the conservation of the generation quantum number g.

The second major development of the GM took place during the period 2003-2011, resulting in a composite model of the colorless leptons and the colored quarks of the SM. Indeed, during the late 20th century, numerous such models had been proposed. The underlying reason for this was that the twelve elementary particles of the SM, the six leptons and the six quarks, was considered to be too many basic building blocks.

The resultant composite 2011 GM was based upon the unified classification scheme and the 1979 models of Harari and Shupe, which were essentially equivalent, employing only two basic massless spin-$\frac{1}{2}$ building blocks (named rishons by Harari): (i) a T-rishon with electric charge $Q = +\frac{1}{3}$ and (ii) a V-rishon with electric charge $Q = 0$.

In order to maintain the pattern of the first generation of leptons and quarks for the second and third generations, a third U-rishon with electric charge $Q = 0$ was employed in the 2011 GM. It was found that the pattern of the first generation could be maintained, if one and two colorless rishon-antirishon pairs, $\Pi = [(\bar{U}V + \bar{V}U]/\sqrt{2}$, were added to the rishon structures of the first generation to form the second and third generations,respectively. Chapter 6 describes the 2011 GM in detail.

In the SM, the color charged quarks are bound within colorless hadrons by the strong chromodynamic force. This raised the question: What force binds the rishons and/or antirishons within the leptons and quarks of the SM?

During 1980-82 several composite models, based upon the 1979 model of Harari and Shupe, involved two color-type $SU(3)$ local gauge theories, namely $SU(3)_H \times SU(3)_C$ at the rishon level. First, one had an additional strong hypergluon interaction, corresponding to the $SU(3)_H$ symmetry, mediated by massless hypergluons, which was responsible for binding rishons, carrying a hypercolor charge, together to form hypercolorless leptons and quarks. This implied that each lepton and quark was required to be composed of three rishons. Second, one retained the strong chromodynamic force, corresponding to the $SU(3)_C$ symmetry, mediated by massless gluons, which is responsible for binding color charged quarks together to form colorless hadrons. However, none of these complicated models provided a satisfactory understanding of the three generations of the SM. I suspect that the two-fold color-charge complication, together with the considerable difficulty involved in relating these models to the nonunified classification scheme of the SM, probably caused the proposed composite models to flounder without achieving any significant success.

In Chapter 7, the origin of mass is discussed. The GM provides a unified origin of mass: the mass of a body m is a measure of its energy E content and is given by $m = E/c^2$, where c is the speed of light in a vacuum, according to Einstein's 1905 conclusion, deduced from his special theory of relativity. Consequently, unlike the SM, the GM has no requirement for a

Higgs field to generate the mass of any particle. Furthermore, if a particle has mass, it is composite.

In Chapter 8, the solution to the enigma of gravity is presented. In the GM, gravity is not a fundamental force but is a very weak, universal and attractive complex residual interfermion color force acting between the colorless particles: electrons, neutrons and protons that essentially constitute the total mass of a body of ordinary matter. This complex residual color force provides a quantum theory of gravity.

Chapter 8 also describes the notions of dark matter and dark energy that the SMC claims constitute about 27% and 68%, respectively, of the mass-energy content of the Universe. However, both dark matter and dark energy are simply names describing unknown entities and consequently constitute two very dubious assumptions of the SMC. In the GM both dark matter and dark energy are understood in terms of properties of the new quantum theory of gravity. First, the nature of the gravitational force in the GM is such that there exists a galactic distance beyond which the gravitational field behaves essentially as $1/r$, rather than as $1/r^2$ as in Newtonian gravity, thereby explaining both the flat rotation curves of spiral galaxies and also the correlation of their asymptotic orbital velocity with their luminosity, known as the Tully-Fisher relation. Second, another property of the gravitational force in the GM is that it has a finite range, estimated to be very approximately 10 billion light years, so that there is no gravitational attraction for very large cosmological distances, thereby offering an explanation of the 1998-1999 observations of Perlmutter, Riess, Schmidt and their collaborators, without requiring the existence of dark energy.

In Chapter 8 I also concluded that the photon interacts with ordinary matter via the new gravitational interaction of the GM, based upon residual color interactions, in agreement with observation, if the photon is assumed to be the standard singlet state of the corresponding QCD color octet of hypergluons binding together the rishons and antirishons of the leptons and quarks in the GM. The singlet hypergluon state, consisting of all three color charges as well as all three anticolor charges, is massless, electrically neutral, colorless and has spin-1 and $U(1)$ symmetry, so that it has all the appropriate properties of the photon. The GM predicts that such photons will be deflected by massive bodies, such as the Sun, by an amount that is twice the value predicted by the GM for the deflection of ordinary matter.

Thus, the gravitational interaction of the GM, predicts the same deflection of light rays, consisting of photons, assumed to be singlet state hypergluons, in agreement with Einstein's general theory of relativity but without any warping of spacetime. It should also be noted that in principle, the gravitational interaction of the GM is also able to account for the anomalous advance of the perihelion of Mercury in terms of the additional term in the gravitational interaction.

Chapter 9 discusses the matter-antimatter asymmetry problem, i.e. the virtual absence of antimatter in the Universe. It is concluded that the SM is incapable of providing an acceptable explanation for the matter-antimatter asymmetry, which is not expected according to the assumed Big Bang scenario, creating the Universe with an equal number of particles and antiparticles from pure energy.

In the GM, it is expected that initially the Big Bang produces numerous rishon-antirishon pairs corresponding to the two kinds of rishons. Then as the Universe cooled, the rishons and antirishons would form leptons, quarks and their antiparticles and eventually atoms of hydrogen, antihydrogen, helium and antihelium. Later, these would annihilate in pairs until only atoms of hydrogen and helium prevailed.

In the GM, it was found essential to define the matter/antimatter nature of composite particles carefully in terms of particle number p. In the GM, the rishons and antirishons have $p = +\frac{1}{3}$ and $p = -\frac{1}{3}$, respectively. Consequently, the initial state of the Universe following the Big Bang has $p = 0$. Since this quantum number is *conserved* in all subsequent interactions, the GM predicts that the present state of the Universe should also have $p = 0$, i.e. no asymmetry. Thus there is *no particle-antiparticle asymmetry* in the present Universe.

It should be noted that the allocation of a finite value of p, i.e. $p = +\frac{1}{3}$ to massless rishons and $p = -\frac{1}{3}$ to massless antirishons, implies that the quantum number p represents mass-energy, rather than pure mass, in addition to its particle or antiparticle nature. Indeed, conservation of p means that mass-energy, or since mass is essentially concentrated energy according to $m = E/c^2$, simply that energy is conserved in the Universe.

The implication of no particle-antiparticle asymmetry in the present Universe, is that the original antirishons (antimatter) created in the Big Bang are now contained within the stable composite leptons, i.e. electrons and neutrinos, and the stable quarks, i.e. the weak eigenstate u-quarks and d'-quarks comprising protons and neutrons. The hydrogen, helium and heavier atoms, following the fusion processes in stars, all consist of electrons, protons and neutrons. This explains where all the original antiparticles have gone, excluding those that annihilated back into pure energy with their corresponding particle. Thus there is *no* matter-antimatter asymmetry in the present Universe. However, it does not explain why the Universe consists primarily of hydrogen atoms and not antihydrogen atoms.

In the GM, an antihydrogen atom consists of the same equal numbers of rishons and antirishons as a hydrogen atom, although the rishons and antirishons are differently arranged in the two systems. This implies that both hydrogen atoms and antihydrogen atoms should be formed in the aftermath of the Big Bang with about the same probability. In fact, estimates from the CMB data suggest that for every billion hydrogen-antihydrogen pairs, there was just one extra hydrogen atom.

I have proposed that this extremely small difference, one extra hydrogen atom in 10^9 hydrogen-antihydrogen pairs, may arise from small statistical fluctuations associated with the complex many-body processes involved in the formation of either a hydrogen atom or an antihydrogen atom. The uniformity of the Universe, in particular, the lack of antihydrogen throughout the Universe, indicates that such statistical fluctuations took place prior to the 'inflationary period' [4] associated with the Big Bang scenario.

It should be noted that following the annihilation of the hydrogen-antihydrogen pairs that the Universe is expected to be dominated by photons and neutrinos, rather than matter particles. However, it is the relatively small number of matter particles that allows us to exist. It should also be noted that the SM does *not* predict this possibility to occur (see Section 9.3).

Chapter 10 discusses the consequences of the postulate of the GM that hadrons are composed of weak eigenstate quarks rather than mass eigenstate quarks. The main consequences are the existence of higher generation quarks in nucleons and the existence of mixed parity states in hadrons. In particular the mixed parity of the charged pions leads to a quantitative description of the decay of the long-lived K_2^0 meson into two charged pions *without the violation of CP symmetry* in the CC weak nuclear interaction process. This is contrary to the earlier interpretation of the surprising discovery by Cronin, Fitch and collaborators in 1964 of the unexpected decay of the K_2^0 meson into two charged pions. Thus the conclusion from this observation is that CP is *conserved* in CC weak nuclear interactions, in agreement with Landau's earlier suggestion.

In Chapter 11 the electroweak connection has been derived within the framework of the GM, by removing the dubious assumption of the SM that the CC weak nuclear interaction is *fundamental*. Thus in the GM, the electroweak connection is derived by employing a $U(1)_p \times SU(2)_L$ *global* gauge transformation, corresponding to the conservation of both particle number p and weak charge g_w. Furthermore, the left-handed nature of the $SU(2)$ symmetry is predicted by the GM as a consequence of the elementary rishons and antirishons being *massless*.

The rishon structures of both the W and Z weak bosons have also been discussed in Chapter 11. In particular, the W bosons have been found to have negative parity, which describes the approximately 100% parity violations observed at low energies of the CC weak nuclear interaction, acting between the six known weak isospin doublets, whose members have the same intrinsic parity. It is suggested that the Z^0 boson consists of both a colorless set of three rishons and a colorless set of three antirishons, in which one of the V-rishons or one of the $\bar{V}$-antirishons exists in an excited 2s state. This structure model suggests the possible existence of additional excited Z^0-type bosons, in which the excited V-rishon or the $\bar{V}$-antirishon exists in a 2p state. Indeed, the boson discovered at CERN with a mass of about 125 GeV with spin and parity 0^+ may be such a boson.

Bibliography

[1] P.A.R. Ade *et al.* (Planck Collaboration), "Planck 2013 results. I. Overview of products and scientific results", Astron. and Astrophys. **571** Art. A1 (2014).

[2] K. Gottfried and V. F. Weisskopf, *Concepts of Particle Physics*, Vol. 1 (Oxford University Press, New York, 1984).

[3] M. J. G. Veltman, *Facts and Mysteries in Elementary Particle Physics*, (World Scientific, Singapore, 2003).

[4] D. Lincoln, *Understanding the Universe from Quarks to the Cosmos*, rev. edn. (World Scientific, Singapore, 2012).

[5] B. A. Robson, "The Generation Model of Particle Physics" in *Particle Physics*, Ed. E. Kennedy (InTech Open Access Publisher, Rijeka, Croatia, 2012).

[6] A. Pais, *Inward Bound: Of Matter and Forces in the Physical World*, (Oxford University Press, New York, 1986).

[7] Y. Ne'eman and Y. Kirsh, *The Particle Hunters*, (Cambridge University Press, Cambridge, 1983).

[8] J. C. Maxwell, "A Dynamical Theory of the Electromagnetic Field" Philosophical Transactions of the Royal Society of London. **155**, 459-512 (1865).

[9] D. C. Peaslee, *Elements of Atomic Physics*, (Prentice-Hall Inc., New York, 1955).

[10] H. A. Lorentz, *The Theory of Electrons* (The Columbia University Press, New York, 1909).

[11] A. Einstein, "Zur Elektrodynamik bewegter Körper", Annalen der Physik. **17**, 891-931 (1905).

[12] A. S. Eddington, *The Mathematical Theory of Relativity*, 3rd edn. (Chelsea Publishing Company, New York, 1975).

[13] J. V. Narliker, *The Lighter Side of Gravity* 2nd edn. (Cambridge University Press, Cambridge, 1996).

[14] H. C. Ohanian, *Einstein's Mistakes* (W.W. Norton and Company, New York, 2009).

[15] S. Tomonaga, *The Story of Spin*, Translated by T. Oka (The University of Chicago Press, Chicago, 1997).

[16] B. A. Robson, *The Theory of Polarization Phenomena* (Clarendon Press, Oxford, 1974).

[17] H. Friedrich, *Theoretical Atomic Physics* (Springer-Verlag, Berlin, Heidelberg, 1990).

[18] S. Hawking, *A Brief History of Time*, (Bantam Press, London, 1988).

[19] M. Gell-Mann and Y. Ne'eman, *The Eightfold Way* (Benjamin, New York, 1964).

[20] F. Wilczek, "In Search of Symmetry Lost", Nature **433**, 239-247 (2005).

[21] M-H. Shao, N. Wang and Z-F. Gao, "Tired Light denies the Big Bang", in *Redefining Standard Model Cosmology*, Ed. B. A. Robson (Intech Open Access Publisher, London, U.K., 2019).

[22] I. J. R. Aitchison and A. J. G. Hey, *Gauge Theories in Particle Physics* (Adam Hilger Ltd, Bristol, 1982).

[23] Y. Fukuda *et al.* (Super-Kamiokande Collaboration), "Measurement of a small atmospheric ν_μ/ν_e ratio", Physics Letters B **433**, 9-18 (1998).

[24] K. Gottfried and V. F. Weisskopf, *Concepts of Particle Physics*, Vol. 2 (Oxford University Press, New York, 1984).

[25] B. A. Robson, "The Generation Model and the Electroweak Connection", International Journal of Modern Physics E **17**, 1015-1030 (2008).

[26] B. A. Robson, "Progressing Beyond the Standard Model", Advances in High Energy Physics **2013**, Art. ID 341738, 12pp.

[27] B. A. Robson, "Generation Model of Particle Physics with Excited Rishon States", Advances in High Energy Physics, Gravitation and Cosmology, **5**, 140-148 (2019).

[28] B. A. Robson, "A Generation Model of the Fundamental Particles", International Journal of Modern Physics E **11**, 555-566 (2002).

[29] I. A. D'Souza and C. S. Kalman, *Preons: Models of Leptons, Quarks and Gauge Bosons as Composite Objects*, (World Scientific, Singapore,1992).

[30] F. Halzen and A. D. Martin, *Quarks and Leptons: An Introductory Course in Modern Particle Physics*, (John Wiley and Sons, New York, 1984).

[31] M. Born, *Atomic Physics*, 2nd Ed., (Blackie and Sons Ltd, London, 1937).

[32] S. Rainville *et al.*, "A Direct Test of $E = mc^2$", Nature **438**, 1096-1097 (2005).

[33] H. Lyre, "Does the Higgs Mechanism Exist", International Studies in the Philosophy of Science **22**, 119-133 (2008).

[34] R. H. Sanders, "Mass Discrepancies in Galaxies: Dark Matter and Alternatives", The Astronomy and Astrophysics Review **2**, 1-28 (1990).

[35] D. Clowe *et al.*, "A Direct Empirical Proof of the Existence of Dark Matter", Astrophysical Journal **648**, L109-L113 (2006).

[36] G. R. Farrer and M. E. Shaposhnikov, "Baryon Asymmetry of the Universe in the Minimal Standard Model", Physical Review Letters **70**, 2833-2836 (1993).

[37] J. H. Christenson, J. W. Cronin, V. L Fitch and R. Turlay, "Evidence for the 2π Decay of the K_2^0 Meson", Physical Review Letters **13**, 138-140 (1964).

[38] A. D. Morrison and B. A. Robson, "2π Decay of the K_L^0 Meson without CP Violation", International Journal of Modern Physics E **18**, 1825-1830 (2009).

[39] E. Abouzaid *et al.* (KTeV Collaboration), "Determination of the Parity of the Neutral Pion via its Four-Electron Decay", Physical Review Letters **100**, 182001 (2008).

Name Index

Subject Index

Printed in the United States
by Baker & Taylor Publisher Services